Flavio Cesar Cunha Galeazzo

Simulation of Turbulent Flows with and without Combustion with Emphasis on the Impact of Coherent Structures on the Turbulent Mixing

Simulation of Turbulent Flows with and without Combustion with Emphasis on the Impact of Coherent Structures on the Turbulent Mixing

by
Flavio Cesar Cunha Galeazzo

KIT Scientific Publishing

Dissertation, Karlsruher Institut für Technologie (KIT)
Fakultät für Chemieingenieurwesen und Verfahrenstechnik, 2013

Tag der mündlichen Prüfung: 18. Oktober 2013
Referenten: Prof. Dr.-Ing. Nikolaos Zarzalis, Prof. Dr. Thomas Wetzel

Impressum

KIT Scientific Publishing

Karlsruher Institut für Technologie (KIT)
KIT Scientific Publishing
Straße am Forum 2
D-76131 Karlsruhe

KIT Scientific Publishing is a registered trademark of Karlsruhe Institute of Technology. Reprint using the book cover is not allowed.

www.ksp.kit.edu

Print on Demand 2016

ISBN 978-3-7315-0408-5
DOI 10.5445/KSP/1000047969

Simulation of Turbulent Flows with and without Combustion with Emphasis on the Impact of Coherent Structures on the Turbulent Mixing

zur Erlangung des akademischen Grades eines

DOKTORS DER INGENIEURWISSENSCHAFTEN (Dr.-Ing.)

von der Fakultät für Chemieingenieurwesen und Verfahrenstechnik des

Karlsruher Instituts für Technologie (KIT)

genehmigte

DISSERTATION

von

M.Sc. Flavio Cesar Cunha Galeazzo

aus São Paulo, Brasilien

Hauptreferent: Prof. Dr.-Ing. Nikolaos Zarzalis

Korreferent: Prof. Dr. Thomas Wetzel

Tag der mündlichen Prüfung: 18.10.2013

Vorwort

Diese Arbeit entstand während meiner Tätigkeit am Lehrstuhl für Verbrennungstechnik des Engler-Bunte-Instituts des Karlsruher Instituts für Technologie (KIT). Die finanzielle Förderung erfolgte durch ein Stipendium des *Conselho Nacional de Pesquisa Tecnológica e Científica* (CNPq) aus Brasilien und durch das Projekt *Theoretische Erfassung der Abhebehöhe von Strahlflammen bei Druckverbrennung* im Rahmen der Forschungsinitiative Kraftwerke des 21. Jahhunderts (KW21), gefördert vom Bundesland Baden-Württemberg und Siemens AG.

Mein besonderer Dank gebührt Herrn Prof. Dr.-Ing. Nikolaos Zarzalis für die Ermöglichung dieser Arbeit und die sehr hilfsbereite und kontinuierliche Betreuung. Für die Übernahme des Korreferates und das freundliche Interesse an dieser Arbeit danke ich Herrn Prof. Dr. Thomas Wetzel.

Die freundschaftliche Atmosphäre am Lehrstuhl für Verbrennungstechnik wird mein ganzes Leben in meiner Erinnerung bleiben. Dabei möchte ich besonders die Numerikgruppe der Verbrennungstechnik hervorheben, stellvertretend N. Sebbar, J. Denev, O. Brunn, T. Voigt, M. Schmitt, I. Dinkov, P. Parthasarathy. Bei der Zusammenarbeit mit M. Kern habe ich besonders viel gelernt. Einem besonderen Dank widme ich die Projektkollegen G. Donnert, T. Lischer, C. Cárdenas und C. Prathap. Vielen Dank auch an der wissenschaftliche Hilfskraft und Kollege C. Bedoya. Allen voraus steht jedoch Dr.-Ing. P. Habisreuther, Leiter der Numerikgruppe. Sein Wissen in Rechnerfragen und die vielen Diskussionen, Ideen und Anregungen waren entscheidend für diese Dissertation.

Dr.-Ing. M. Hettel danke ich ganz herzlich für die zahlreichen Diskussionen, für die ganz breite Unterstützung und für das Korrekturlesen dieser Arbeit.

Herrn W. Pfeffinger danke ich sehr, der durch seine unermütliche Arbeit einen reibungslosen Ablauf der Forschung am Institut ermöglicht. Mit ihm habe ich viel über wissenschaftliche Rechen gelernt.

Für die umfassende, herzliche Unterstützung, bedanke ich mich bei meiner alten Freunde Daniela, Alexandra und Marcelo. Schließlich gilt mein weitaus größter Dank meinen Eltern, Maria Inez und Enivaldo, und meinem Bruder, Rafael Bruno, für ihre Unterstützung, Ermutigung und Verständnis. Imenso é o amor que tenho por vocês.

Karlsruhe, im Oktober 2013 *Flavio Cesar Cunha Galeazzo*

Kurzfassung

Die Berechnung der turbulenten Mischung in komplexen turbulenten Strömungen ist eine anspruchsvolle Aufgabe. Der auf verschiedenen Längen- und Zeitskalen bis in den molekularen Bereich hinein ablaufende Mischungsprozess hängt wesentlich von der Dynamik innerhalb des turbulenten Strömungsfeldes ab. Der einfachste Fall der turbulenten Mischung findet sich zwischen passiven Skalaren. Beispiele sind Mischungen von Flüssigkeiten mit der gleichen Dichte oder Fluide, welche mit nicht reagierenden Markierungssubstanzen beladen sind. Es gibt jedoch auch Strömungen, bei denen die turbulente Mischung die Fluideigenschaften selbst ändert. Dies ist der Fall, wenn die Strömung durch zusätzlich ablaufende chemische Reaktionen überlagert wird, wie z. B. bei Verbrennungprozessen.

Die turbulente Mischung ist ein dreistufiger Prozess der sich durch die Vorgänge des *Entrainments* (grobskalige Einmischung durch kohärente Strukturen), des *Rührens* (makroskopische Vermischung durch nahezu richtungsinvariante Turbulenzwirbel) und der molekularen Diffusion zusammensetzt. Die numerische Modellierung des turbulenten Mischungsprozesses muss diese drei Phänomene berücksichtigen. Während das *Rühren* und die Diffusion in der Regel durch Turbulenzmodelle erfasst werden, bleibt die Beschreibung des Entrainments außerhalb Ihres Leistungsbereiches. Der Grund dafür ist, dass das Entrainment in turbulenten Scherströmungen vor allem von den großen kohärenten Strukturen geprägt wird. Diese sind jedoch nicht im ursprünglichen Sinne turbulent und abhängig von der äußeren Strömungsgeometrie. Für derartige Strömungen müssen instationäre Berechnungen unter Verwendung von Reynolds gemittelten Navier-Stokes (RANS) oder der Large Eddy Simulation (LES) Methoden verwendet werden, welche die zeitliche Entwicklung der kohärenten Strukturen direkt erfassen können, damit der Einfluss des Entrainments prinzipiell berücksichtigt wird.

Zunächst wurde der in der Literatur gut dokumentierte Fall eines turbulenten konzentrischen Freistrahles berechnet, um die verschiedenen numerischen Werkzeuge und Simulationsmethoden zu testen und zu validieren. Sowohl instationäre als auch stationäre Simulationen konnten wichtige Merkmale der Freistrahlkonfiguration mit guter Genauigkeit vorhersagen. Es zeigte sich jedoch eine systematische Abweichung zwischen der unter Verwendung der stationären RANS Simulation berechneten Mischung in der Scherschicht und den entsprechenden Messergebnissen. Die in dieser Zone auftretenden Wirbelringe, die durch die Kelvin-Helmholtz Instabilität entstehen, können nur mit der Methode der LES vorhergesagt werden, was zu einer besseren Übereinstimmung mit den Messungdaten führt.

Der Jet in Crossflow (JIC) ist eine weit komplexere Strömungskonfiguration als der konzentrische Freistrahl. Es treten hierbei verschiedene Typen von kohärenten Strukturen sowie Rezirkulationsgebiete auf. Die Qualität der Übereinstimmung der Simulationsergebnisse mit den Messungen ist daher stark an die Qualität der Beschreibung der kohärenten Strukturen gekoppelt. Die Ergebnisse der LES Simulationen geben die diese Strukturen und deren Auswirkungen viel detaillierter wieder als die Ergebnisse einer instationären RANS Rechnung. Dagegen ist eine stationäre RANS Simulation nicht in der Lage kohärente Strukturen abzubilden. Damit werden deren Auswirkungen komplett vernachlässigt. Die Übereinstimmung der Ergebnisse der LES mit den Messungswerten ist sehr gut. Die mittels instationären RANS Rechnungen erzielten Ergebnisse zeigen dagegen eine schlechtere Übereinstimmung. Die schlechtesten Ergebnisse liefern Rechnungen, die mit der Technik der stationären RANS durchgeführt wurden. Während die Übereinstimmung der mittels stationärer RANS berechneten Geschwindigkeitsfelder gegenüber den Messwerten noch relativ befriedigend ist, wird der turbulente Mischungsprozess in Strömungsbereichen, die durch das Auftreten von kohärenten Strukturen dominiert sind, stark unterschätzt. Dieser scheinbare Widerspruch konnte durch die Verwendung des Konzeptes der Intermittenz geklärt werden, mit welchem sich die unterschiedlichen Wirkungen der kohärenten Strukturen auf Geschwindigkeits- und Skalarfelder beschreiben lassen.

Die numerischen Methoden, die für isotherme Strömungen angewendet wurden, wurden zur Simulation eines komplexen Verbrennungssystems erweitert und angepasst, bei dem zwei wichtige Quellen zur Ausbildung von Instationaritäten vorhanden sind: die Entwicklung von großen kohärenten Strukturen und die sich im Injektor ausbildenden akustischen Resonanzen. Diese instationären Phänomene spielen eine tragende Rolle für die Charakteristik der turbulenten Mischung und damit auch für die Flammengeometrie. Dies hat wiederum Auswirkungen auf die Temperatur- und Speziesverteilungen und beeinflusst unter anderem die Vorhersage der Schadstoffbildung. Beide instationären Phänomene konnten mit den LES Simulationen wiedergegeben werden und liefern insgesamt eine gute Übereinstimmung der Rechenergebnisse mit den Messergebnissen. Die stationäre RANS Simulationen vernachlässigen beide Quellen instationären Verhaltens und zeigen somit eine deutlich schlechtere Übereinstimmung mit der Messergebnisse. So wird unter anderem die Flammengeometrie völlig unbefriedigend wiedergegeben. Die Schlussfolgerung ist, dass zeitaufgelöste Berechnungsmethoden wie die LES erforderlich sind, um derart komplexe reagierende Strömungen mit guter Genauigkeit zu modellieren.

Die Ergebnisse haben gezeigt, dass stationäre RANS Simulationen eine gute quantitative und qualitative Übereinstimmung mit den experimentellen Daten zeigen, wenn der Einfluss von großen kohärenten Strukturen auf die mittlere Strömungsgeschwindigkeit vernachlässigbar ist. Wenn jedoch die mittlere Strömungsgeschwindigkeit durch kohärente Strukturen oder andere Quellen beeinflusst wird und zeitliche Schwankungen aufweist, so stimmen die Reynolds-Mittelwerte nicht mit den Zeit-Mittelwerten überein. In diesem Fall muss eine instationäre Berechnungsmethode verwendet werden. Wenn jedoch die Strömung nicht statistisch stationär ist, d. h. mit ausgeprägten Schwankungen in der mittleren Strömungsgeschwindigkeit durch große kohärente Strukturen oder andere Quellen behaftet ist, so stimmen die Reynolds-Mittelwerte nicht mit den Zeit-Mittelwerten überein. Da diese Schwankungen ursprünglich nicht turbulenter Natur sind, wird deren Einfluss auf die mittlere Strömung nicht von Turbulenzmodellen erfasst. Um aussagekräftige Ergebnisse zu erzielen, sind in diesem Fall zeitaufgelöste Simulationen zwingend notwendig. Die zu bevorzugende Methode ist die LES, welche die kohärenten Strukturen weitaus detaillierter erfasst als eine instationäre RANS Rechnung. Die vorgelegte Arbeit zeigt die Grenzen der Methode der stationären RANS auf und dokumentiert die Notwendigkeit von instationären Rechnungen für die untersuchten Strömungskonfigurationen. Die verantwortlichen Gründe und Phänomene werden aufgezeigt, detailliert analysiert und deren Wirkungen anschaulich dargestellt.

Abstract

The analysis of turbulent mixing in complex turbulent flows is a challenging task. The effective mixing of entrained fluids to a molecular level is a vital part of the dynamics of turbulent flows, especially when combustion is involved. The simplest case of turbulent mixing occurs between passive scalars. Examples are mixing of fluids with the same density or loaded with non-reacting tracers. There are cases, however, in that the turbulent mixing changes the fluid characteristics. Combustion phenomena are remarkable examples of such systems with active scalars.

Turbulent mixing is a three-stage process of entrainment, stirring and diffusion. The numerical simulation of turbulent mixing has to consider these three different phenomena. While the stirring and the diffusion are usually modeled together by turbulence models, the entrainment remains out of their scope. The reason is that in turbulent free shear flows, the entrainment is promoted mainly by the large-scale coherent structures typical for these flows. These coherent structures are not turbulent in nature and highly geometry and time dependent. In this situation, high-level simulation techniques such as unsteady Reynolds-averaged Navier Stokes (RANS) or large eddy simulation (LES) have to be used, which can resolve the temporal progress of coherent structures directly, naturally accounting for the influence of the entrainment.

The free jet, a flow configuration extensively studied in the literature, has been used to test and validate the numerical tools and the different simulation methodologies. Both unsteady and steady state simulations have been able to predict important features of the free jet with good accuracy. There is, however, a systematic discrepancy between the steady state RANS results and the measurements of the mixing near the jet edge. Only the LES could predict the vortex rings originated by the Kelvin-Helmholtz instability present in this region, resulting in a better agreement with the measurements.

The jet in crossflow (JIC) is a more complex flow configuration than the free jet, with a series of large-scale coherent structures and recirculation regions. In the JIC, the quality of the agreement of the simulation results with the measurements is strongly coupled with the description of the coherent structures. The LES reproduced the coherent structures in much more detail than the unsteady RANS, while the steady state RANS simulations do not resolve the coherent structures at all, neglecting this important source of unsteadiness. The agreement of the LES results with the measurements is very good, with worse agreement using unsteady RANS and steady state RANS simulations, in this order. The results of the RANS simulations have shown less intense mixing in the flow regions dominated by coherent structures, even though the agreement of the velocity fields with the measurements is good. It was shown that this fact can be explained applying the concept of the intermittency, which determines the different influence of the coherent structures on the velocity and passive scalar fields.

The numerical tools used for isothermal flows were extended and adapted to the simulation of a complex combustion system, in which two important sources of unsteadiness are present: the development of large-scale coherent structures and an acoustic resonance in the injector region. These time-dependent phenomena play a major role in the turbulent mixing, affecting directly the flame geometry and consequently the temperature and species distributions, with severe consequences for the prediction of pollutant formation, for example. The LES simulations predicted both time-dependent phenomena, producing results that agreed well with the measurements. The steady state RANS simulations neglected these important sources of unsteadiness, resulting in a far less satisfactory agreement especially for the flame geometry. The conclusion is that time-resolved computational methods are required to model such complex reacting flows with good accuracy.

The results have shown that steady state RANS simulations provide good quantitative and qualitative agreement with experimental data when the flow is statistically stationary, *i.e.*, when the influence of large-scale coherent structures or unsteadiness of the mean flow are negligible. However, in flows with pronounced unsteadiness in the mean flow promoted by large-scale coherent structures or other sources, Reynolds-averaged values do not converge to their time-averages. As this unsteadiness is not turbulent in nature, its influence on the mean flow is not modeled by the turbulence models. Hence, to achieve high-fidelity results, time dependent simulations are mandatory, which increases the computational cost substantially. The preferred method is the LES, which resolved the coherent structures in much more detail than the unsteady RANS. The work has shown the limitations of the steady-state RANS simulations and acknowledged the need of applying unsteady methods for the calculation of the investigated flow configurations. The responsible reasons and phenomena have been identified, analyzed in detail and their impact clearly illustrated.

Contents

Nomenclature

Roman	units	
a	m/s	speed of sound
b_i	-	one-dimensional low-pass filter
B_u, B_c	-	constants of the free jet
$B_{i,j,k}$	-	three-dimensional low-pass filter
c	-	passive scalar, reaction progress variable
C	-	mean passive scalar
$c_{p,i}$	-	specific heat capacity at constant pressure of species i
d	-, m	variance of the reaction progress variable, distance
D	m^2/s^2,m	thermal or molecular diffusivity, diameter
D_k	$kg/m\cdot s^2$	equivalent diffusion coefficient of species k into the mixture
f	Hz, -	characteristic frequency, mixture fraction
$f_{k,j}$	m^2/s^2	volume force acting on species k in direction j
g	-	variance of the mixture fraction
G	-	filter function
h	J/kg	specific enthalpy
h_k^0	J/kg	specific enthalpy of formation of species k
I	-	intermittency function
J_u	m^3/s^3	diffusion flux of velocity
J_c	m^2/s^2	diffusion flux of passive scalar

Roman	units	
k	m^2/s^2, –	turbulent kinetic energy per unit mass, complex wave number
l, ℓ, L	m	characteristic length
$\mathscr{L}_{ij}$	m^2/s^2	Germano identity
n	-	local coordinate normal to the surface, normalized length scale
N	-	number of realizations
p	Pa	pressure
P	Pa, -	mean pressure, probability
q_j	m/s	subgrid scale flux vector
r^α	-	random fields
R	-	jet-to-crossflow velocity ratio
$R_{I,J,K}$	-	filtered fields
R_{uu}	-	autocorrelation function
Re	-	Reynolds number
s_{ij}	1/s	strain rate tensor
S_{ij}	1/s	mean strain rate tensor
S	1/s, m^2, –	modulus of the mean strain rate tensor, area, spreading rate
t	s	time
t_{ij}	$kg/m\cdot s^2$	viscous stress tensor
T	s, K	mean time, temperature
u_i; u, v, w	m/s	velocity vector components
U_i; U, V, W	m/s	mean velocity vector components
U_b	m/s	characteristic speed
$\overline{u'_i u'_j}$	m^2/s^2	Reynolds stress tensor for incompressible flows
$\overline{\rho}\widetilde{u''_i u''_j}$	$kg/m\cdot s^2$	Reynolds stress tensor for compressible flows
$\overline{\rho}\widetilde{u''_i Y''_k}$	$kg/m^2\cdot s$	Reynolds flux vector for mass transfer in compressible flows
$\overline{\rho}\widetilde{u''_i h''}$	kg/s^3	Reynolds flux vector for heat transfer in compressible flows
$\overline{u'_i c'}$	m/s	Reynolds flux vector for incompressible flows

Roman	units	
V	m^3	volume
$V_{k,i}$	m/s	i-component of the diffusion velocity V_k
x_i; x, y, z	m	Cartesian coordinate components
x, r, θ	m	cylindrical coordinate components
Y_k	-	mass fraction of species k
Z_k	-	mass fraction of atom k

Greek	units	
δ	m	characteristic thickness
δ_{ij}	-	Kronecker delta
Δ	m	filter width
ε	m^2/s^3	dissipation rate of k per unit mass
η	-	non-dimensional radial coordinate
λ	W/m·K	thermal conductivity
ω	1/s, m/s, Hz	specific dissipation rate of k, outgoing pressure wave, frequency
$\dot{\omega}_k$	$kg/m^3 \cdot s$	reaction rate of species k
ρ	kg/m^3	density
ϕ	-	generic scalar, equivalence ratio
ψ	N/m^2	compressibility
ν	m^2/s^2	kinematic viscosity
ν_t	m^2/s^2	eddy viscosity in incompressible flows
μ	$kg/m \cdot s^2$	dynamic viscosity
μ_t	$kg/m \cdot s^2$	eddy viscosity in compressible flows
γ	-	intermittency factor, blending factor
Γ	-	gamma function
σ	-	Prandtl / Schmidt number
σ_t	-	turbulent Prandtl / Schmidt number
τ_W	N/m^2	wall shear stress
τ_{ij}	N/m^2	subgrid scale stresses
ξ	-	relaxation factor

Superscript	
ϕ'	fluctuation, subgrid scale part
ϕ''	unsteadiness about the mean value, Favre fluctuation
$\overline{\phi''^2}$, $\widetilde{\phi''^2}$	variance
ϕ^+	non-dimensional value in wall units

Subscript	
ϕ_0	reference value
$\phi_{1/2}$	half-radius value
ϕ_{inf}	value at infinity
ϕ_c	value conditioned on the crossflow, centerline value
ϕ_{cross}	mean crossflow value
ϕ_{Eff}	effective (turbulent plus laminar) value
ϕ_F	fuel stream
ϕ_j	value conditioned on the jet flow
ϕ_{jet}	mean jet value
ϕ_n	value conditioned on the non-turbulent ambient flow
ϕ_{Ox}	oxidizer stream
ϕ_{ref}	reference value
ϕ_{rms}	root-mean-square value
ϕ_t	turbulent, value conditioned on the turbulent jet flow

Operator	
$\overline{\phi}$	Reynolds averaged / filtered value
$\widetilde{\phi}$	Favre, mass weighted averaged / filtered value
$\langle\phi\rangle_t$	conditional average
$\langle\phi\rangle_T$	time average
$\langle\phi\rangle_S$	spatial average
$\langle\phi\rangle_E$	ensemble average

Abbreviations	
BC	boundary condition
CFD	computational fluid dynamics
DES	detached eddy simulation
DNS	direct numerical simulation
EBI-VBT	Engler-Bunte-Institute, Division of Combustion Technology
FGM	flamelet generated manifolds
ICCD	intensified charge-coupled device
JIC	jet in crossflow
JPDF	joint probability density function
KH	Kelvin-Helmholtz
LDA	laser Doppler anemometry
LES	large eddy simulation
LIF	laser-induced fluorescence
MPI	message passing interface
PFR	plug flow reactor
PIV	particle image velocimetry
RANS	Reynolds-averaged Navier Stokes
rms, RMS	root-mean-square
SAS	scale-adaptive simulation
SGS	subgrid scale
SST	shear-stress tensor RANS turbulence model
URANS	unsteady Reynolds-averaged Navier Stokes
USST	unsteady RANS simulation with SST turbulence model
UV	ultraviolet

List of Figures

List of Tables

1 Introduction

The effective mixing of entrained fluids to a molecular level is a vital part of the dynamics of turbulent flows, especially when combustion is involved. The motions induced from the largest to the smallest eddies transport fluid in the turbulent region, greatly enlarging the interfacial surface area between them, which in turn enhances the overall mixing rate. Following the terminology of Eckart [25], turbulent mixing can be viewed as a three-stage process of entrainment, stirring, and diffusion. Entrainment is the movement of one fluid caused by another, as in the engulfment of ambient fluid by a jet. Stirring is defined as the mechanical process of distributing fluids more uniformly in a given domain, increasing their interfacial area, while mixing is the diffusion process of substances across interfacial surfaces. Mixing is a molecular process, depending on material properties such as diffusivities, while entrainment and stirring are purely kinematical processes that depend on flow parameters.

The simplest case of turbulent mixing occurs between passive scalars. Examples are mixing of fluids with the same density or loaded with non-reacting tracers. In these systems, a correct description of the mixing is not required by the flow dynamics as the mixing of these fluids has a negligible back effect on the flow. There are cases, however, in that the turbulent mixing changes the fluid characteristics, e.g., its composition, density or enthalpy, which can in turn have a significant impact on the flow dynamics. Combustion phenomena are remarkable examples of such systems with active scalars. In particular, the buoyancy-driven flow that both sustain and is driven by a candle is a familiar example.

Simulation of mixing in turbulent flows is of great importance from theoretical and practical points of view. The even increasing computational resources allow the simulation of larger, more complex systems in shorter timeframes. The dissemination of this technology comes with new challenges, as the increased pressure on the robustness, accuracy and predictive capabilities of the models, which are faced with more complex systems and phenomena.

Turbulence models are one of the main building blocks of modern Computational Fluid Dynamics (CFD). Nevertheless, the modeling of complex turbulent flows requires much more than the simple modeling of the underlying turbulence. Turbulent flows with a heavy presence of large-scale structures have additional sources of unsteadiness than simpler turbulent flows, which requires different strategies for their modeling.

One methodology to account for the influence of large-scale structures in turbulent flows is the intermittency, which began with the pioneer work of Townsend [112] in 1948 and was object of intense studies in the 1970s and beginning of 1980s. The experimental apparatus of choice was the hot wire anemometer, which could easily be coupled with a cold wire system to measure time-resolved, correlated velocity-temperature data in flows with a small temperature difference. The focus on the intermittency decreased in the 1980s with the popularization of Laser Doppler Anemometry (LDA) for velocity measurements, which made the acquisition of correlated velocity-concentration data more difficult. At almost the same time, steady state CFD simulation using Reynolds-averaged Navier-Stokes (RANS) turbulence models became popular, which are a good match to LDA measurements.

With the current advance in the measurement and simulation technique, the study of turbulent mixing has gained new momentum. The newly developed high-resolution measurements using simultaneous Particle Image Velocimetry and Laser Induced Fluorescence of jets in crossflow realized at the Engler-Bunte-Institute [67, 30, 31, 18], for example, show the new possibilities of laser-based, non-intrusive measurement technique that is able to deliver high quality correlated velocity-scalar data. On the other hand, unsteady simulations using RANS or large eddy simulation (LES) techniques have become increasingly popular with the advance in computing power. The new measurements enabled the quantification of the known weaknesses of steady state simulations in cases where large-scale structures dominate the flow.

Returning to the terminology discussed in the first paragraph, the simulation of turbulent mixing has to consider three different phenomena: entrainment, stirring and diffusion. In flows with heavy influence of large-scale structures, the effects of the entrainment cannot be neglected, which increases the interest in unsteady simulations. The intermittency proved itself an appropriate framework to account for the influence of these phenomena on turbulent flows, and was employed in the discussion of the measurements and simulations in this work. With this tool, much of the observed erratic behavior of the steady state simulations of turbulent mixing of passive scalars can be explained.

Turbulent mixing is a complex phenomenon itself, which becomes even more complex in reacting flows. Most of the scalars in reacting flows are active, with a two-way interaction between chemistry and turbulence. On one hand, the active character of the scalars increases the need of a correct description of the turbulent mixing. On the other hand, the presence of large-scale structures complicates the description of the same turbulent mixing. To solve this riddle, the tools developed for unsteady simulations of turbulent flows were applied to turbulent combustion, significantly improving the description of turbulent mixing in the presence of large-scale structures.

1.1 Objectives

The primary objective of this work is to develop simulation tools capable of describing the turbulent mixing in complex turbulent flows with and without combustion.

One of the main tasks is to comprehend the reason why some steady state RANS simulations of passive scalar mixing produce results in good agreement with the measurements, while some have a poor agreement using the same framework. After careful analysis, it becomes clear that the description of the turbulent mixing was the responsible for the inaccuracy of the steady state RANS simulations.

In turbulent flows without combustion, the study of turbulent mixing is facilitated by the use of passive scalars. For some systems, like the free jet, steady state RANS simulations are satisfactory even using very basic models for the turbulent mixing. However, for other systems of relevance as the jet in crossflow, while the velocity field is adequately represented, the inaccuracy of the predicted mixing field is high.

A careful analysis pointed out that the problem lies in large-scale coherent structures, which belong to almost all turbulent free shear flows. The influence of coherent structures over the mean field is not modeled by RANS turbulence models. The coherent structures are highly geometry dependent, which makes the derivation of a general model of their effect a complex task. Fortunately, unsteady simulations using the RANS or LES frameworks can solve the coherent structures directly, accounting for their influence over the time-averaged field.

Numerical tools capable of simulating mixing in turbulent flows with a high level of accuracy have been developed. The tools and the methodology have been tested and validated using a free jet configuration, which is a test case extensively studied in the literature. The same tools and methodology were then applied on the jet in crossflow configuration, exposing the similarities and differences of the two flows.

The knowledge gained by the study of turbulent mixing is employed in the simulation of a complex reacting system, in which the turbulent mixing plays a major role in the flame geometry. The numerical tools developed for non-reacting flows were adapted for turbulent combustion, and specific models for the interaction between turbulence and chemical reactions were applied. Steady state RANS simulations and unsteady LES were conducted, and the results were compared to experimental data.

1.2 Outline

The thesis is divided into two main parts. The first part deals with the theories, derivations and methodologies employed as part of the investigation, and spans from Chapter 2 to 5. The second part focuses on the validation of the simulation approaches using a series of test cases from Chapters 6 to 8. Overviews of each chapter are presented below.

The theoretical background of turbulent flows and turbulent mixing is introduced in Chapter 2, with a description of turbulence, its origin, nature and statistical treatment. The theory behind turbulent reacting flows is also developed in this chapter.

Chapter 3 is devoted to coherent structures, their origin and nature. The choice of a new chapter dedicated to coherent structures is deliberate, to state clearly the difference between them and other phenomena present in turbulent flows.

The knowledge of Chapters 2 and 3 flows into Chapter 4, which discusses the phenomenon of intermittency and its consequences for the description of turbulent flows.

Closing the first part of the thesis, Chapter 5 introduces the numerical tools used in the simulations. The turbulent models used in the simulations are presented, along with details about the numerical methods. In addition, the turbulent reaction models used in the simulation of combustion systems are introduced.

The second part of the thesis begins with the free jet presented in Chapter 6, a flow configuration extensively studied in the literature, which has been used to test and validate the numerical tools and the different simulation methodologies. The chapter begins with a short review of the literature data and continues with the numerical setup used in the simulations. The results of the simulations are compared to experimental data and discussed.

In Chapter 7, the ideas and tools developed earlier are applied to the jet in crossflow configuration. This flow configuration is of major theoretical and practical importance. It is more complex than the free jet, with a series of large-scale coherent structures and recirculation regions. As the previous chapter, it begins with a review of the experimental investigations and follows with the numerical setup used in the simulations. The simulations are carefully compared with experimental data, supporting the discussion of the results through the chapter.

The simulation tools developed and tested in the free jet (Chapter 6) and jet in crossflow (Chapter 7) configurations have been applied to the High Strain Burner, described in Chapter 8. In this chapter, these tools are employed in the simulation of a complex combustion system, in which the turbulent mixing plays a crucial role in the flame geometry and consequently in the whole system. The results of different simulation methodologies are compared with experimental data.

The thesis is concluded in Chapter 9.

2 Turbulent Flows

This chapter contains a brief description of the fundamental physical processes that govern turbulence and turbulent mixing in incompressible flows, along with the mathematical equations that in turn govern them. In addition, the statistical treatment of the governing equations, which is essential for the study of turbulent flows, can be found in this chapter. The text continues with the equations used to describe turbulent reacting flows.

2.1 Fundamental equations

An uncontroversial fact about turbulent flows is that they are the most complex kind of flow. Despite their complexity, turbulent flows are still described by the Navier-Stokes equations, which express the conservation of mass and momentum[1] for a flow of continuum, Newtonian fluid. Although the linear relation between the viscous stresses and the rate of strain in Newtonian fluids being the simplest applicable to this problem, it follows that the solution of the equations for turbulent flows lead to a chaotic behavior, limiting the direct solution of the equations to very specific cases.

As even the smallest turbulent eddies have a size much larger than the mean free path of the gas molecules unless the Mach number (velocity of fluid to the speed of sound in that medium) is extremely high, the continuum condition of the Navier-Stokes equations is valid for turbulent flows [93]. Similarly, the constitutive equations of the most common gases and liquids follows the linear Newtonian viscous-stress law. The Navier-Stokes equations are therefore adequate to study the low Mach number, turbulent gaseous flows in this work.

[1] In the modern literature, the Navier-Stokes equations refer to the system of equations encompassing the conservation of mass and momentum. In old textbooks the Navier-Stokes equation referred only to the conservation of momentum.

The Navier-Stokes system of equations for conservation of mass and momentum for an incompressible, Newtonian fluid can be written as

$$\frac{\partial u_i}{\partial x_i} = 0 \tag{2.1}$$

$$\frac{\partial u_i}{\partial t} + \frac{\partial u_i u_j}{\partial x_j} = -\frac{1}{\rho}\frac{\partial p}{\partial x_j} + \frac{\partial}{\partial x_j}\left(\nu \frac{\partial u_i}{\partial x_j}\right) \tag{2.2}$$

where x_i and u_i are the space and velocity vectors with i=(1,2,3), t is the time, ρ is the density, p is the pressure and ν is the kinematic viscosity. Following the Einstein summation convention, repeated indices mean summation over the index. The derivation of these equations can be found in many fluid mechanics textbooks [3, 4].

In order to study the mixing, the concept of a passive scalar is introduced. The adjective *passive* means that the fluid elements carrying the scalar are convected and diffused by molecular motion without altering the velocity field in which they coexist. In contrast, if a flow exhibits large density differences due to the mixing of fluids with different densities, for example, there is a dynamic coupling between the velocity and scalar fields. The scalar is not passive in this case.

The conservation equation of a passive scalar c is given by

$$\frac{\partial c}{\partial t} + \frac{\partial u_j c}{\partial x_j} = \frac{\partial}{\partial x_j}\left(D \frac{\partial c}{\partial x_j}\right) \tag{2.3}$$

where D is the molecular diffusivity, considering that c represents the concentration of a trace species.

An important characteristic of the passive scalar is its *boundedness*. If the initial and boundary conditions of c lie within a given range

$$c_{min} \leq c \leq c_{max} \tag{2.4}$$

then c must lie in this range in the whole domain for any given time: values greater of c_{max} or less than c_{min} cannot occur.

The similarity between Equation (2.2) without the pressure gradient and Equation (2.3) form the base for the analogy between momentum and scalar (heat and mass) transfer. Unfortunately, along with the inherent differences between the transport of vector and scalar quantities, the pressure gradient prevents an exact analogy in turbulent flows, as pressure fluctuations always accompany velocity fluctuations [7]. This fact will be used in the discussion about the different effects of intermittency in velocity and mixing fields in Chapter 4.

2.2 The origin of turbulence

The Reynolds number, defined in Equation 2.9, represents the ratio of inertial forces to viscous forces. In flows that are originally laminar, turbulence arises from instabilities associated with large Reynolds numbers. Flows with a large Reynolds number are dominated by inertial forces; the viscous forces are not able to smooth out the flow instabilities. Laminar pipe flow becomes turbulent at a Reynolds number of about 2300, based on the mean velocity and diameter. Boundary layers without pressure gradient become unstable at a Reynolds number of 600, based on the displacement thickness and the free stream velocity. Free shear flows become unstable at very low Reynolds numbers because of an inviscid instability mechanism that does not operate in boundary layers and pipe flows [110].

On the other hand, as dictated by its dissipative nature, turbulence cannot maintain itself but depends on the energy supply of the environment. A common source of energy for the turbulence is shear in the mean flow, as can be seen in Chapter 5, especially Equation (5.2). If turbulence arrives in an environment where no shear or other maintenance mechanism is present, it decays. The turbulent Reynolds number decreases and the flow tends to become laminar again. The decay of the turbulence created in a uniform flow that flows through a grid is a classic example of this phenomenon.

Turbulence is always related to high Reynolds number flows and originates itself from the non-linear convective term of the Navier-Stokes equation. Using the scaling parameters in Table 2.1, the variables of Equation (2.2) have been nondimensionalized

$$t^* = ft \ , \ x_i^* = x_i/l \ , \ u_i^* = u_i/U_b \ , \ p^* = \frac{p - p_\infty}{p_0 - p_\infty} \ , \ \frac{\partial}{\partial x_i^*} = l\frac{\partial}{\partial x_i}. \quad (2.5)$$

Substituting these variables in the Equation 2.2 and multiplying by l/U_b^2 results in the following equation[2]

$$\left[\frac{fl}{U_b}\right]\frac{\partial u_i^*}{\partial t^*}+\frac{\partial u_i^* u_j^*}{\partial x_j^*}=-\left[\frac{p_0-p_\infty}{\rho U_b^2}\right]\frac{\partial p^*}{\partial x_j^*}+\left[\frac{\nu}{U_b l}\right]\frac{\partial}{\partial x_j^*}\left(\nu\frac{\partial u_i^*}{\partial x_j^*}\right), \quad (2.6)$$

where three non-dimensional numbers can be identified:

$$\frac{fl}{U_b}=\text{St (Strouhal number)}, \quad (2.7)$$

$$\frac{p_0-p_\infty}{\rho U_b^2}=\text{Eu (Euler number)}, \quad (2.8)$$

$$\frac{U_b l}{\nu}=\text{Re (Reynolds number)}. \quad (2.9)$$

In a flow with increasing Reynolds number, the relative importance of the diffusion term decreases, as it scales with 1/Re. The non-linear convection term becomes the dominant term of the equation, originating the turbulent fluctuations when the stabilizing effect of the diffusion term decreases.

Scaling parameters	Description	Dimension
U_b	Characteristic speed	m/s
l	Characteristic length	m
f	Characteristic frequency	1/s
$p_0 - p_\infty$	Reference pressure difference	kg/m·s^2

Table 2.1: Scaling parameters.

2.3 The nature of turbulence

While an intuitive idea about the nature of turbulent flows is part of our everyday life, it is however very difficult to define it precisely. Hinze [39] defined turbulent fluid motion as "an irregular condition of the flow in which the various quantities show a random variation with time and space coordinates,

[2] Equation 2.6 is nondimensionalized, but not normalized. To be normalized, all nondimensional variables have to span approximately between zero and unity, which depends heavily on the choice of the scaling parameter.

so that statistically distinct averaged values can be discerned". Many textbooks avoid giving turbulence a clear definition, which can be sometimes misleading. Other authors as Tennekes and Lumley [110] and Libby [63], instead of a definition, opted to cite some characteristics of turbulent flows. Some of these characteristics are listed below.

As turbulent flows always show *irregular* or *random* variations, a deterministic solution of turbulence problems is not possible; statistical methods are used instead.

The increased rates of momentum, heat and mass transfer and the rapid mixing associated with turbulent flows are caused by the increased *diffusivity* of turbulence. This is the single most important feature as far as applications are concerned. Examples span from the prevention of boundary layer separation on airfoils at large angle-of-attack, to increased heat and mass transfer rates in processes of all kinds, and to increased energy density of turbulent flames.

Turbulence always occurs at flow with *high Reynolds number*. Turbulence often originates as an instability of laminar flows; these instabilities are related to the interaction of viscous terms and nonlinear inertial terms in the governing equations. This interaction is described by the Reynolds number, which can be understood as the ratio of the nonlinear inertia terms and the viscous terms in the governing equations. As seen in the previous section, when the Reynolds number increases, the viscous terms lose their importance and the inertia terms dominate, which in turn originates the irregular turbulent motion.

Turbulence is *rotational* and *three dimensional*. Flows that are essentially two dimensional, such as cyclones in the atmosphere or vortex streets after cylinders are not turbulent themselves, even though their characteristics may be strongly influenced by small scale turbulence generated somewhere by shear or buoyancy, which interacts with the larger flow.

Another characteristic of turbulent flows is that they are always *dissipative*. The deformation work performed by the viscous shear stresses increases the internal energy of the fluid at the expense of kinetic energy of the turbulence. The dissipative nature of the turbulence dictates that it needs a continuous supply of energy to compensate the viscous losses; without it the turbulence decays rapidly.

The phenomenon of turbulence occurs in *continuum* fluids, governed by the equations of fluid dynamics. Even the smallest scales of turbulence are far larger than any molecular length scale.

Turbulence is not a feature of fluids, but of fluid flows: *turbulent flows are flows*. The major characteristics of turbulent flows are not controlled by the molecular properties of the fluid, what makes the dynamics of turbulence the same for all fluids if the Reynolds number is large enough, even though these characteristics depend on the environment. This dependency on the environmental conditions prevents a general way to deal with all kinds and types of turbulent flows. The theory concentrates on families of flows with simple boundary conditions, like boundary layers, jets and wakes.

The fluctuations in turbulent flows always involve a *wide range of scales*, with the large scales having a relative permanence and the small scales having relatively short lifetimes. Indeed, turbulence is distinguished from unsteady laminar flow in terms of the continuous spectrum of scales of the fluctuations involved. The cascade process arises from this theory, in which the energy from the large-scale fluctuation is transferred into the smaller scales that are directly affected by the molecularity.

2.4 Statistical treatment of turbulent flows

The Navier-Stokes equations and the conservation equation of the passive scalar, Equations (2.1) to (2.3), combined with suitable boundary conditions, are sufficient for the simulation of laminar flows with the transport of a passive scalar. The simulation of turbulent flows based only on these equations is called direct numerical simulation (DNS), and no modeling[3] is required. As all scales of the flow have to be resolved, the DNS becomes increasingly computationally expensive as the Reynolds number increases and the turbulent time and length scales decrease. An example of the increasing prohibitive computational resources required for a DNS is shown in Annex A.1.

Turbulent flows exhibit a *random*[4] variation of its quantities [121, 93]. Fortunately, virtually all engineering applications are not interested in the instantaneous value of the different variables, but in their statistics like mean values. A statistical treatment of turbulent flows can produce exactly this kind of information, while avoiding the burden of a DNS.

[3] Except the model for the molecular viscosity and diffusivity.

[4] The meaning of a random variable is that it does not have a unique value - the same every time the experiment is repeated under the same set of conditions. The Navier-Stokes equations remain naturally valid.

O. Reynolds [99] introduced the procedure of expressing the turbulent quantities as the sum of a mean and a fluctuating part. The decomposition of the velocity $u_i(x_i,t)$ in its mean $U_i(x_i,t)$ and fluctuating parts $u_i'(x_i,t)$

$$u_i(x_i,t) = U_i(x_i,t) + u_i'(x_i,t) \tag{2.10}$$

is referred to as Reynolds decomposition. The Reynolds average, in its more general version, is represented by an overbar, *i.e.*, the mean value of u_i is $\overline{u_i} \equiv U_i$.

Three forms of averaging are the most pertinent to the turbulence modeling: *time average*, *spatial average* and *ensemble average*. The general term used to describe these averaging processes is *mean*. The time average is appropriate for *statistically stationary* turbulent flows, *i.e.*, a turbulent flow whose averaged quantities do not vary with time, as a fully developed pipe flow driven by a constant pressure difference. The time average operation $\langle\ \rangle_T$ is defined by

$$U_i(x_i) = \langle u_i(x_i,t)\rangle_T = \lim_{T\to\infty} \frac{1}{T}\int_t^{t+T} u_i(x_i,t)dt. \tag{2.11}$$

An illustration of this process can be seen in Figure 2.1. The solid line represents the instantaneous values, while the dotted line represents the time average.

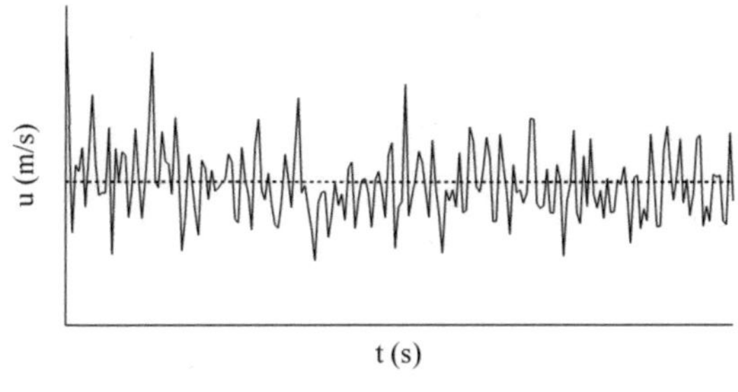

Figure 2.1: Time averaging for a statistically stationary flow.

For homogeneous turbulence, which is on average uniform in all directions, the spatial averaging is appropriate. All spatial coordinates are covered by a volume integral, and the averaged $\langle\,\rangle_S$ is defined by

$$U_i(t) = \langle u_i(x_i,t)\rangle_S = \lim_{V\to\infty}\frac{1}{V}\int_V u_i(x_i,t)dV. \tag{2.12}$$

The most general form of Reynolds averaging is the ensemble average $\langle\,\rangle_E$, which is suitable for flows that are neither statistically stationary nor homogeneous. For example, flows with decay in time or with large-scale structures. In an idealized example, the ensemble average is defined in terms of N identical realizations of the same experiment, with initial and boundary conditions that differ only by random infinitesimal perturbations

$$U_i(x_i,t) = \langle u_i(x_i,t)\rangle_E = \lim_{N\to\infty}\frac{1}{N}\sum_{n=1}^{N} u_i(x_i,t). \tag{2.13}$$

Figure 2.2 illustrates this concept. The ensemble average is represented by the dashed line. The instantaneous values and the time average are represented by the solid and dotted lines, respectively.

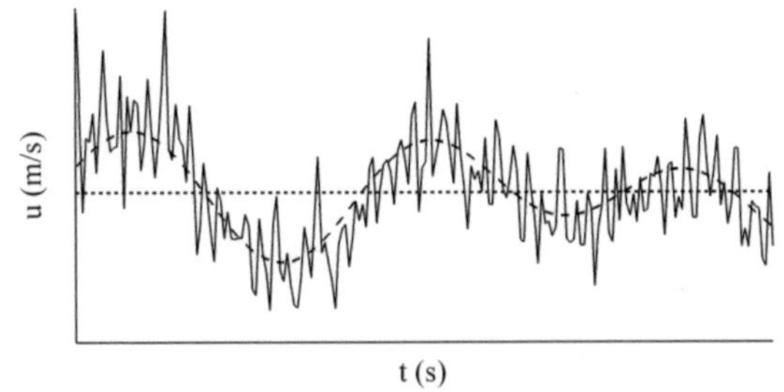

Figure 2.2: Ensemble and time averaging for a statistically non-stationary flow.

At this point, it is essential to note the difference between the time and ensemble averages. Considering a statistically stationary flow, the time averaged velocity is defined by Equation (2.11). The time average of the mean velocity is again the same mean velocity, *i.e.*,

$$\langle U_i(x_i)\rangle_T = \lim_{T\to\infty}\frac{1}{T}\int_t^{t+T} U_i(x_i)dt = U_i(x_i). \tag{2.14}$$

Considering now a flow that is statistically non-stationary, the mean flow contains slow variations in time that are not turbulent in nature. Flows with large-scale structures caused by vortex shedding or a duct flow with an imposed slowly varying periodic pressure gradient are good examples. The ensemble-average is the appropriate averaging method in these cases, and the mean velocity $U_i(x_i,t)$ can be time-dependent. It follows that a time average can still be defined; however the time-average mean velocity is different from the mean velocity

$$\langle U_i(x_i,t)\rangle_T = \lim_{T\to\infty}\frac{1}{T}\int_t^{t+T} U_i(x_i,t)dt = U_i(x_i) \neq U_i(x_i,t). \tag{2.15}$$

Figure 2.2 illustrates this difference. The values of the ensemble average and the time average are indeed decidedly different. This finding leads to the definition of the so called triple decomposition. In this case, the turbulent fluctuations about the ensemble-average $u_i'(x_i,t)$ and the unsteadiness of the mean flow about the time-average $u_i''(x_i,t) = U_i(x_i,t) - U_i(x_i)$ are treated separately

$$u_i(x_i,t) = U_i(x_i) + u_i'(x_i,t) + u_i''(x_i,t). \tag{2.16}$$

The triple decomposition has little value in practice, as often the boundary between the unsteadiness of the mean flow and the turbulent fluctuations is not very distinct, making the evaluation of the term $u_i''(x_i,t)$ difficult or impossible. The implications of time-averaging a statistically non-stationary flow are, however, of great significance. When measuring instantaneous velocity components in a statistically non-stationary flow and evaluating the time average applying Equation (2.11), the corresponding fluctuations about the mean are the sum of the turbulent fluctuations about the ensemble-average $u_i'(x_i,t)$ and the unsteadiness of the mean flow about the time-average $u_i''(x_i,t)$.

It is important to note that despite the difficulty in separating the turbulent fluctuations from the unsteadiness of the mean flow when analyzing experimental data, the physical phenomena that originates these two fluctuations are quite different. While the turbulent fluctuations are the result of the non-linearity of the convection term of the equation of conservation of momentum (Equation 2.2) when the Reynolds number becomes large, the unsteadiness of the mean flow can be originated from external influences as a varying pressure gradient, or self-generated aero- or hydrodynamical phenomena as vortex streets, which are present also in laminar flow.

2.4.1 Averaging the fundamental equations

The Navier-Stokes equations can be treated using the same Reynolds decomposition and averaging technique. Substituting the instantaneous flow quantities for the sum of mean and fluctuating components and taking the average yields the so-called Reynolds averaged Navier-Stokes (RANS) equations. Being the most general Reynolds average, the ensemble average is used. The equations in Cartesian tensor form for incompressible, Newtonian fluid are

$$\frac{\partial U_i}{\partial x_i} = 0 \tag{2.17}$$

$$\frac{\partial U_i}{\partial t} + \frac{\partial U_i U_j}{\partial x_j} = -\frac{1}{\rho}\frac{\partial P}{\partial x_i} + \frac{\partial}{\partial x_j}\left(\nu \frac{\partial U_i}{\partial x_j}\right) - \frac{\partial \overline{u_i' u_j'}}{\partial x_j} \tag{2.18}$$

where the overbar is shorthand for the Reynolds average. The process of averaging the equations is illustrated, for example, in references [121, 93].

They have the same general form as the instantaneous Navier-Stokes equations, with the substitution of the instantaneous variables by the mean variables, and the inclusion of the last term $-\partial\overline{u_i' u_j'}/\partial x_j$ in the momentum equation. Replacing Equation (2.17) in Equation (2.1) reveals that the fluctuation velocity u_i' also has zero divergence. The quantity $\rho\overline{u_i' u_j'}$ is known as *Reynolds stress tensor*, so that $\overline{u_i' u_j'}$ is the *specific Reynolds stress tensor* which must be determined in order to solve Equation (2.18).

The Reynolds-averaged equation for the conservation of a passive scalar can be derived in a similar way:

$$\frac{\partial C}{\partial t} + \frac{\partial U_j C}{\partial x_j} = \frac{\partial}{\partial x_j}\left(D\frac{\partial C}{\partial x_j}\right) - \frac{\partial \overline{u_i' c'}}{\partial x_j} \tag{2.19}$$

where the *specific Reynolds flux vector* $\overline{u_i' c'}$ must also be determined in order to solve Equation (2.19).

The influence of the turbulence on the mean flow is contained in the Reynolds stresses and fluxes. Conservation equations for them can be derived from the fundamental equations, but they contain again more unknowns, which are a consequence of the non-linearity of the Navier-Stokes equation: every time the variables are decomposed and the equations are averaged, new unknowns are created. This is called the *closure problem.* At the point of view of the physics of the problem, it is not an unexpected result, as the Reynolds averaging does not add any new physical principle to the system. As stated by Wilcox [121]

> In essence, Reynolds averaging is a brutal simplification that loses much of the information contained in the Navier-Stokes equation.

The role of the turbulence modeling (see Section 5.2) is to provide approximations for the Reynolds stresses and fluxes, derived in terms of known quantities, in a way to close the system of equations.

2.5 Reacting flows

Poinsot and Veynante [91] highlight the three main differences between the incompressible, non-reacting flow described in the last sections and a reacting flow:

- a reacting gas is a non-isothermal mixture of multiple species (hydrocarbons, oxygen, water, carbon dioxide, etc.) which must be tracked individually. Thermodynamic data are also more complex, changing significantly with temperature and composition,
- species react chemically and the reaction rate requires specific modeling,
- since the gas is a mixture, transport coefficients require specific attention.

The exothermic reaction of a fuel and a oxidant is also called combustion. Combustion can occur in either a flame or a nonflame mode, and flames, in turn, are categorized as being either premixed flames or nonpremixed (diffusion) flames [114]. A thin zone of intense chemical reaction is what commonly characterizes a flame. If the reaction takes place simultaneously in many locations within a volume, a nonflame combustion occurs, also named *flameless oxidation*.

The two classes of flames (premixed and nonpremixed) are related to the state of mixedness of the reactants. In a premixed flame, the fuel and the oxidizer are mixed at the molecular level before any relevant reaction. The spark-ignition engine works with premixed flames. On the other hand, in a nonpremixed (diffusion) flame the reactants are initially separated and the reaction occurs only at the interface between them. A candle is an example of a diffusion flame. In practical devices, it is often difficult to classify the flame as pure premixed or pure diffusion. The complex combustion system which will be presented in Chapter 8 shares characteristics of both premixed and diffusion flames.

To characterize the quality of a fuel-air mixture, the equivalence ratio ϕ can be used. If exactly enough oxidizer is available to react completely with the fuel, the ratio is known as the stoichiometric mixture. The equivalence ratio is defined as the ratio of the fuel-to-oxidizer ratio to the stoichiometric fuel-to-oxidizer ratio

$$\phi = \frac{Y_{fuel}/Y_{oxid}}{(Y_{fuel}/Y_{oxid})_{stoich}} \tag{2.20}$$

where Y represents the mass fraction and the suffix *stoich* stands for stoichiometric conditions.

2.5.1 Fundamental equations for reacting flows

The derivation of the fundamental equations for the transport of mass, species and energy is found in textbooks as those by Williams [122] and Kuo [56]. This section concentrates on the forms used in the simulations presented in Chapter 8.

Overall continuity is given by:

$$\frac{\partial \rho}{\partial t} + \frac{\partial \rho u_i}{\partial x_i} = 0. \tag{2.21}$$

The mass conservation equation for the species k is written:

$$\frac{\partial \rho Y_k}{\partial t} + \frac{\partial \rho u_i Y_k}{\partial x_i} = -\frac{\partial V_{k,i} Y_k}{\partial x_i} + \dot{\omega}_k, \quad \text{for } k = 1, N. \tag{2.22}$$

where $V_{k,i}$ is the i-component of the diffusion velocity V_k of species k and $\dot{\omega}_k$ is the reaction rate of species k. By definition:

$$\sum_{k=1}^{N} V_{k,i} Y_k = 0 \text{ and } \sum_{k=1}^{N} \dot{\omega}_k = 0. \tag{2.23}$$

The rigorous evaluation of V_k is difficult and costly. The Hirschfelder and Curtis [40] approximation is used instead:

$$V_k X_k = -D_k \frac{\partial X_k}{\partial x_i} \text{ with } D_k = \frac{1 - Y_k}{\sum_{j \neq k} X_j / D_{jk}}. \tag{2.24}$$

The coefficient D_k is not a binary diffusion but an equivalent diffusion coefficient of species k in the rest of the mixture. This approximation is convenient because the diffusion coefficients D_k can be simply linked to the heat diffusivity D_{th} in many flames, as the Lewis number of individual species $\mathrm{Le}_k = D_{th}/D_k$ usually varies by small amounts in flame fronts.

The equation of momentum is given by:

$$\frac{\partial \rho u_j}{\partial t} + \frac{\partial \rho u_i u_j}{\partial x_i} = -\frac{\partial p}{\partial x_j} + \frac{\partial}{\partial x_i}\left(\mu \frac{\partial u_i}{\partial x_j}\right) + \rho \sum_{k=1}^{N} Y_k f_{k,j}, \tag{2.25}$$

where $f_{k,j}$ is the volume force acting on species k in the direction j. Even though no explicit reaction term is included in this equation, the flow is modified by combustion: the dynamic viscosity μ and the density ρ change strongly because temperature also varies in a ratio from 1:8 to 1:10. The dilatation through the flame front accelerates the flow in the same ratio. Consequently, the Reynolds number varies, leading to very different flow behavior with or without combustion.

The equation of energy requires attention because multiple forms exist. The specific enthalpy h has the advantage of not being modified by chemical reactions, simplifying the equation. The heat release due to chemical reaction only transforms chemical energy into sensible heat.

Considering a mixture of ideal gases, h is evaluated from the specific enthalpies of its components

$$h = \sum_{i}^{N_i} Y_i h_i. \tag{2.26}$$

with h_i given by

$$h_i = h_k^0(T_{ref}) + \int_{T_{ref}}^{T} c_{p,i} dT, \tag{2.27}$$

where h_k^0 is the specific enthalpy of formation of species k at the reference temperature T_{ref} and $c_{p,i}$ is the specific heat capacity at constant pressure of species i.

The transport equation of h is

$$\frac{\partial \rho h}{\partial t} + \frac{\partial \rho u_i h}{\partial x_i} = \frac{\partial}{\partial x_i}\left(\lambda \frac{\partial T}{\partial x_j}\right) + \frac{Dp}{Dt} + \frac{\partial u_i}{\partial x_i}\left(\mu \frac{\partial u_i}{\partial x_j}\right). \tag{2.28}$$

The simplifying assumption of Lewis number equal to unity has been used in the derivation of this equation.

In analogy with the mass conservation equation for species k, Equation (2.22), a transport equation for a scalar c can be formulated using Fick's law of diffusion

$$\frac{\partial \rho c}{\partial t} + \frac{\partial \rho u_i c}{\partial x_i} = \frac{\partial}{\partial x_i}\left(\rho D_c \frac{\partial c}{\partial x_i}\right) + \dot{S}_c \tag{2.29}$$

where D_c is defined as in Equation (2.24) and $\dot{S}_c$ is the source or sink of the scalar. In contrast to the passive scalar defined in Chapter 2, a reacting flow exhibits large density differences due to the heat release, for example, leading to a dynamic coupling between the velocity and scalar fields. The scalar is not passive in this case.

2.5.2 Turbulent reacting flows

When the flow in a flame is turbulent, turbulence and combustion interact. Turbulent combustion is found in most engineering applications, as rockets, internal combustion engines, gas turbines, industrial boilers and furnaces.

In constant density flows, Reynolds averaging is used to derive equations suitable for turbulent flows (see Section 2.4.1). Using this procedure in reacting flows leads to unclosed quantities, for example the correlation between density and velocity fluctuations $\overline{\rho' u_i'}$. To avoid this difficulty, mass-weighted averages, introduced by Favre [26], are preferred

$$\widetilde{f} = \frac{\overline{\rho f}}{\overline{\rho}}. \tag{2.30}$$

Any quantity f can be split into mean and fluctuating components as

$$f = \widetilde{f} + f'' \quad \text{with } \widetilde{f''} = 0. \tag{2.31}$$

Mass-weighted averaging suppresses the terms containing correlations involving density fluctuations. Using this formalism, averaged balance equations formally identical to the Reynolds-averaged equations can be derived [91]:

Overall continuity

$$\frac{\partial \overline{\rho}}{\partial t} + \frac{\partial \overline{\rho}\widetilde{u}_i}{\partial x_i} = 0, \tag{2.32}$$

Momentum

$$\frac{\partial \overline{\rho}\widetilde{u}_j}{\partial t} + \frac{\partial \overline{\rho}\widetilde{u}_i\widetilde{u}_j}{\partial x_i} = -\frac{\partial \overline{p}}{\partial x_j} + \frac{\partial}{\partial x_i}\left(\mu \frac{\partial \widetilde{u}_i}{\partial x_j}\right) - \frac{\partial \overline{\rho}\widetilde{u_i'' u_j''}}{\partial x_j} + \overline{\rho}\sum_{k=1}^{N} \widetilde{Y}_k f_{k,j}, \tag{2.33}$$

Chemical species

$$\frac{\partial \overline{\rho}\widetilde{Y}_k}{\partial t} + \frac{\partial \overline{\rho}\widetilde{u}_i\widetilde{Y}_k}{\partial x_i} = \frac{\partial}{\partial x_i}\left(\overline{\rho}\overline{D}_k \frac{\partial \widetilde{Y}_k}{\partial x_i}\right) - \frac{\partial \overline{\rho}\widetilde{u_i'' Y_k''}}{\partial x_i} + \overline{\dot{\omega}}_k, \quad \text{for } k = 1, N, \tag{2.34}$$

Enthalpy

$$\frac{\partial \overline{\rho}\widetilde{h}}{\partial t} + \frac{\partial \overline{\rho}\widetilde{u}_i\widetilde{h}}{\partial x_i} = \frac{\partial}{\partial x_i}\left(\overline{\lambda}\frac{\partial \widetilde{T}}{\partial x_j}\right) + \overline{\frac{Dp}{Dt}} + \overline{\frac{\partial u_i}{\partial x_i}\left(\mu\frac{\partial u_i}{\partial x_j}\right)} - \frac{\partial \overline{\rho}\widetilde{u_i''h''}}{\partial x_i}. \quad (2.35)$$

The unclosed terms $\overline{\rho}\widetilde{u_i''u_j''}$, $\overline{\rho}\widetilde{u_i''Y_k''}$ and $\overline{\rho}\widetilde{u_i''h''}$ in the equations above are closed using turbulence models, which will be presented in Chapter 5.

The models used to calculate the source term $\overline{\dot{\omega}}_k$ in Equation (2.34), which is actually the main additional problem when dealing with turbulent reacting flows, will also be presented in Chapter 5.

3 Coherent Structures and Mixing

This quotation from Hussain [43] summarizes the main point of this chapter:

> In many cases, the coherent structures are highly dominant. That is, in many cases they are not perturbations of the time-mean flow: they are the flow.

The idea of coherent structures being small perturbations about the mean flow is misleading. For many flows of interest, including the jet in crossflow, coherent structures are so large and energetic that they cannot be considered simply as fluctuations or perturbations.

This chapter begins with the definition of coherent structures. The text continues with a brief discussion about the occurrence of coherent structures in shear flows, followed by the origin and nature of coherent structures. A discussion about stirring and mixing and their relation to coherent structures closes this chapter.

3.1 Definition of coherent structures

A clear definition of coherent structures is still a matter of debate [7, 42, 43]. Even the term *coherent structure* is not generally accepted; some authors prefer *orderly structures* or *inviscid instabilities*.

For the purpose of this work, the term coherent structures will be used to describe large-scale structures with distinct characteristics distinguishing them from turbulence (definition see Chapter 2). For example, all structures that are two-dimensional in nature are not turbulent, and are qualified as coherent structures.

3.2 Coherent structures in shear flows

This section is a short review of the evidence of large-scale, coherent structures in shear flows. It begins with some findings in two-dimensional mixing layers, jets and jets in crossflow. Figure 3.1 shows sketches of the three cases discussed below.

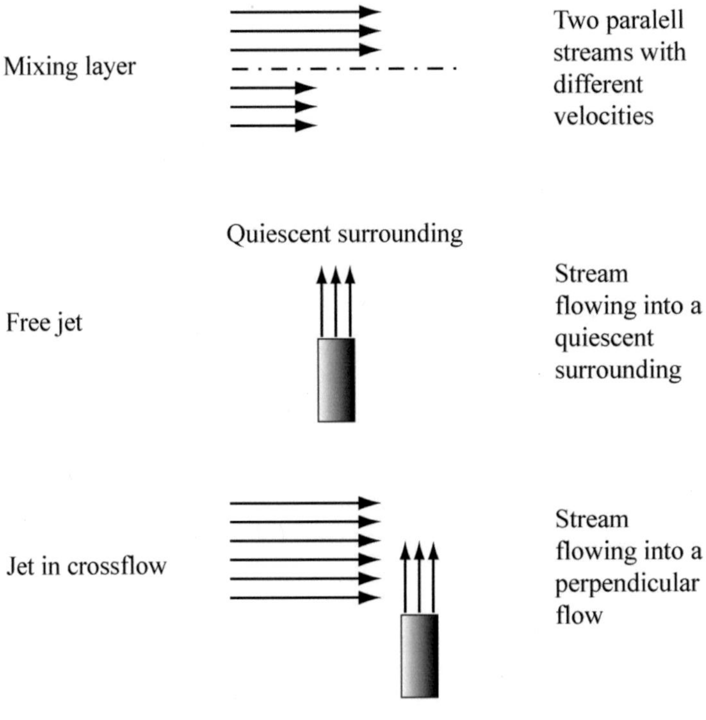

Figure 3.1: Sketches of the mixing layer, jet and jet in crossflow configurations.

Mixing layers

Evidence of the existence of large-scale coherent structures in shear flows has received significant attention over the years. Figure 3.2 shows one of the classic shadowgraph pictures of Brown and Roshko [11] of the plane mixing layer formed between helium and nitrogen flows. Large two-dimensional structures as well as the fine-scale structure, which exist throughout the mixing region, are clearly visible. It is important to note that shadowgraphs are used to make the scalar field visible, not the velocity field. The sharp edges in the visualizations are consistent with the description of scalar fields in intermittent flows, which will be discussed in detail in Chapter 4. The velocity field does not feature such sharp edges, as noted by Broadwell and Mungal [10].

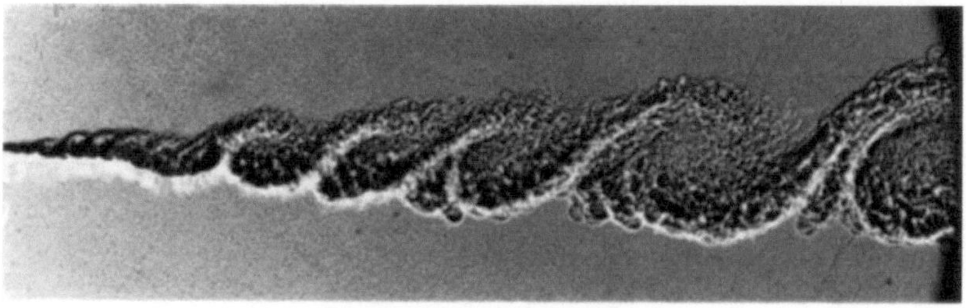

Figure 3.2: Shadowgraph of a high Reynolds number mixing layer. The helium stream on top moves at a velocity of 10 m/s, and the nitrogen on the bottom at a speed of 3.78 m/s. The whole test section is pressurized to 8 bar, giving a Reynolds number based on downstream distance of the order of 10^6. Reproduced from [11].

The classic Kelvin-Helmholtz instability explains the formation of these structures [3]. The Kelvin-Helmholtz instability is present in shear flows, where small-scale perturbations are amplified, drawing kinetic energy of the main flow [15]. This phenomenon is inherently irrotational and two-dimensional, being described by the Euler equations. A more detailed introduction to the concept of the Kelvin-Helmholtz instability can be found in the Annex A.2.

The occurrence of Kelvin-Helmholtz structures in mixing layers is reported to exist from moderate Reynolds numbers of order 20.000 [11] up to high Reynolds numbers exceeding 200.000 [10]. These facts provide compelling evidence that the large-scale structures are essentially inviscid in nature, and exist for all Reynolds numbers beyond some critical level.

Jets

As a consequence of the increased complexity of the jet in comparison to the two-dimensional mixing layer, the evidence of organization in jets is much less direct. However, modern diagnostics and visualization techniques have revealed the underlying large-scale structures of the flow.

Older techniques as shadowgraphs or Schlieren images cannot reveal the organization of the flow, as the structures are three-dimensional and one structure is often nested within others. Laser light sheet techniques as the one applied in Figure 3.3 are more adequate to visualize the large-scale structures of the jet. The Kelvin-Helmholtz instabilities are clearly visible in this figure. In addition, the use of movie sequences and chemical reactions to make structures visible are successful.

Figure 3.3: The Kelvin-Helmholtz shear layer rollup in a round jet, LIF cross section. Reproduced from [116].

Dahm and Dimotakis [19] used a pH-sensitive fluorescent dye and a movie sequence to investigate turbulent water jets at a maximum Reynolds number of 10.000. They have found evidence for the presence of large-scale organization of entrainment and mixing in the self-similar field of the jets. One particularly interesting finding is depicted in Figure 3.4. In contrast to the mean concentration profile, the composition of the mixed fluid is approximately uniform within large regions extending one local jet diameter in both axial and radial directions. Depending on which instant the concentration is sampled across the jet, the profiles have either a top-hat profile, indicating fairly homogeneous concentration across the jet, or a two-level profile, indicating that two distinct regions with different concentrations can be discerned. These regions with fairly homogeneous concentration are the result of the Kelvin-Helmholtz structures, which are convected downstream. Another finding is that the "flame tip" from the movie sequence, *i.e.* the location where the jet fluid fully reacted with the ambient fluid, oscillated in a quasi-periodic fashion. Careful examination of results indicates that this oscillation is a result of the large-scale structures of the flow, which are convected downstream and react quasi-periodically at the tip.

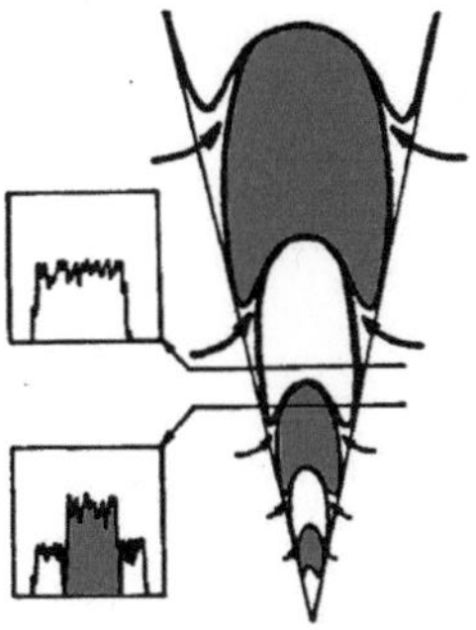

Figure 3.4: Idealized conceptual picture of the instantaneous concentration field of the turbulent jet, showing schematically top-hat (upper) and two-level (lower) profile shapes. Alternating flow structures have been colored gray and white to help to differentiate them. Adapted from [19].

Mungal and Hollingsworth [78] have drawn the same conclusions about the occurrence of large-scale organization in a gaseous jet flow, however using a radically different setup. They analyzed a movie sequence of the exhaust plume of a Titan IV rocket motor, with a Reynolds number of about $2{\cdot}10^8$. They concluded that the large-scale organization is associated with inviscid instability mechanisms that are Reynolds number independent.

Jets in crossflow

The jet in crossflow configuration is the result of the complex, three dimensional interaction between the jet and crossflow streams.

Directly after exiting the nozzle the jet acts as a bluff body, creating a region of high pressure, stagnated flow upstream, and a low pressure region downstream of the jet. The jet quickly bends in the direction of the crossflow (see Figure 3.5). This deflection can be explained by two mechanisms: the pressure gradient formed at the jet exit between the high pressure upstream and the low pressure downstream regions, and the entrainment and mixing of crossflow into the jet flow.

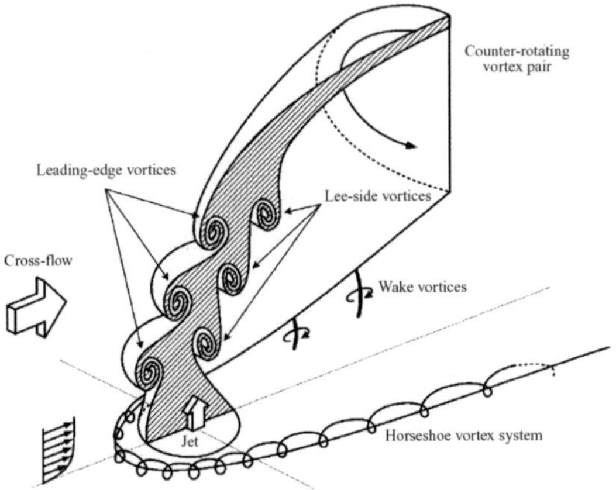

Figure 3.5: Scheme of the jet in crossflow phenomenology. Reproduced from [81].

The large-scale structures observed in the near field are either similar to those originated from the Kelvin-Helmholtz instability in round jets, or to the horseshoe vortices formed at a cylinder-wall junction. The roll-up process of the shear layer happens at both upstream and downstream of the jet, giving origin to the leading-edge vortices and lee-side vortices (lee side is the side that is sheltered from the crossflow). The leading-edge vortices are far more evident and occur over a larger time scale than the lee-side vortices. The reason is that, on the lee side, the vortices break down quickly and the mixing process occur within a short distance. This process can be visualized in Figure 3.6, where colored dye was added to the water flow to make the structures visible. The horseshoe vortices, on the other hand, are very similar to that formed around a cylinder-wall junction.

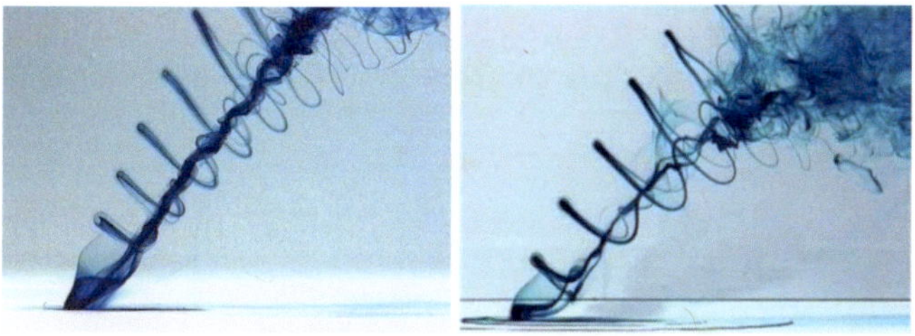

Figure 3.6: Dye pattern of a jet in a crossflow issuing from an elliptical pipe with aspect ratio = 0.3. The dye was released through a dye port located slightly upstream of the jet exit. Velocity ratio R = 3 (left) R = 2 (right). Jet Reynolds number from 900 to 5100. Reproduced from [65], boundary conditions described in [81].

Investigations on the jet in crossflow, focusing on the mixing of chimney plumes, started in the 1930's [83, 68]. Since the pioneer work of Fric & Roshko [28], that visualized the jet in crossflow using a smoke-wire technique, the pseudo-wake that develops downstream of the jet has been another focus of interest. Evidence that alternate vortices are shed in a similar way to the Kármán vortex street formed behind solid bluff bodies is widespread [28, 50, 77, 125]. These wake vortices, also described as upright or tornado-like vortices, are shed to the leeward side of the jet and are connected to the core jet flow, transporting vorticity from the boundary layer to the center [28].

The counter-rotating vortex pair, the most prominent large-scale structure develops further downstream. It is a feature of the mean flow that persists in the far-field [9, 48] and appears even in steady and laminar numerical simulations [16]. It seems that the counter rotating vortex pair will always be present, independent of the velocity ratio, the Reynolds number and the shape of the nozzle, because they are an essential feature of this flow [100]. Using a water channel, Camussi *et al.* [13] have observed the formation of the counter-rotating vortex pair in a jet in crossflow of Reynolds number 100 (see Figure 3.7). Even at this very low Re, the flow is dominated by large-scale structures, as can be seen in the flow visualization depicted in Figure 3.8. Time-resolved measurements and simulations [106, 125] have shown that, while being symmetrical in average, the instantaneous structures that compose the vortex pair are not symmetrical. Rather, the Kármán-like vortices present in the pseudo-wake drive an intense fluctuation of the intensity of the two counter rotating vortices. The counter rotating vortex pair is strongly modulated by the flow field of the coherent structures that, in fact, is as important as the mean velocity field.

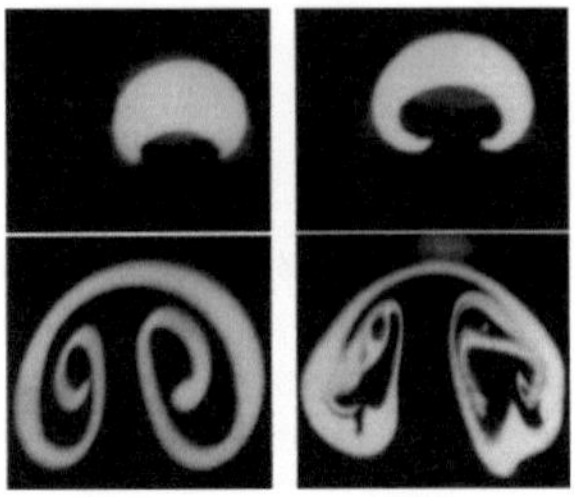

Figure 3.7: LIF visualization of the counter-rotating vortices forming within the jet flow for different distances from the jet exit. The plane of the photograph is perpendicular to the jet axis. Reproduced from [13].

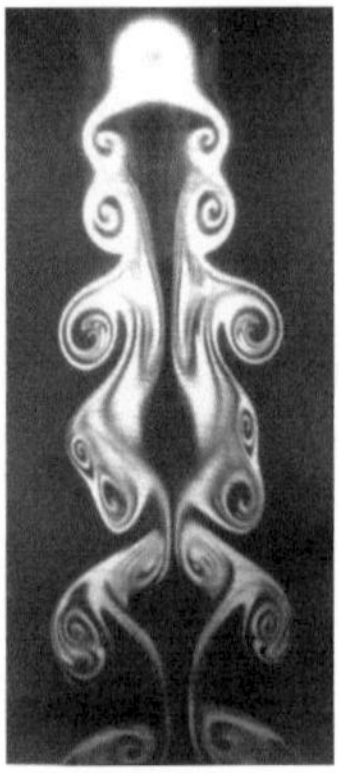

Figure 3.8: LIF visualization obtained by placing the laser sheet parallel and very close to the wall ($x/D = 0.2$, R = 2). Reproduced from [13].

The origin of several of these structures strongly depends on the high pressure gradients developed in the jet exit region, both inside and outside the nozzle. The fast recovery of the pressure field behind the jet is responsible for the vortex breakdown that occur there, resulting in small-scale turbulence production. This phenomenon is illustrated in Figure 3.9. The Kelvin-Helmholtz structures are formed, and the part of these structures that is at the leading-edge side of the jet remain clearly discernible in the whole area of the picture. The part of these structures on the lee side, on the other hand, is broken and mixed, leading to the vortex breakdown.

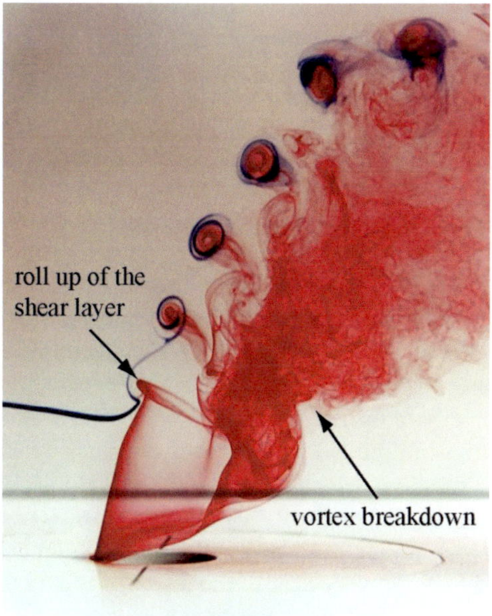

Figure 3.9: Jet in a cross flow. Kelvin-Helmholtz instabilities form on the jet column. Adapted from [65].

The wake or Kármán-like vortices are associated with the alternate shedding in the downstream region of the jet due to the lateral separation of the wall boundary layer, like in the classical von Kármán vortex street. In addition, the deformation of the jet cross section, which yields the counter rotating vortex pair, can be explained in terms of the pressure field around the jet, and the initial acceleration that the jet fluid experiences in the direction of the crossflow [16, 80]. Therefore, the pressure field around the jet exit plays a dominant role in the formation of the large-scale structures of the jet in crossflow [100].

3.3 The nature of coherent structures

The large-scale structures that are present in mixing layers and free jets are attributed to the Kelvin-Helmholtz (KH) instability, illustrated in Figures 3.2 and 3.3. The KH instability is not turbulent itself, as turbulence is always rotational and three-dimensional as discussed in Section 2.3. During its development, the instability can naturally interact with the turbulence and influence mixing phenomena. The interaction is, however, very complex and depends heavily on the boundary conditions of the problem.

The jet in crossflow share characteristics from both free jets and flows around bluff bodies, which is also evident in its large-scale structures. The similarity of the cross sections of a jet in Figure 3.3 and a jet in crossflow in 3.10 is evident, with the Kelvin-Helmholtz instabilities clearly recognizable. In Figure 3.10, the velocity ratio was approximately 4.6 and the jet Reynolds number was about 1600. Kelso *et al.* [50] used flow visualization to study the jet in crossflow in a water channel, with Reynolds numbers based on the jet diameter and crossflow velocity in the range of 440 to 6200. The ring vortices in Figure 3.9 have their origin in the Kelvin-Helmholtz instability at the shear layer between jet and crossflow, which is similar to the KH instability of a free jet and is present even at the smallest Reynolds number. The initiation of the counter rotating vortex pair was also visualized, as the shear layer of the jet is seen to fold and roll up very near to the jet exit, leading to the vortex pair formation. The pressure gradient at the jet exit is responsible for this behavior and for the observed vortex breakdown directly downstream the jet exit.

Figure 3.10: Round jet in a crossflow, LIF cross section. Reproduced from [66].

The pseudo-wake of a jet in crossflow is not a wake in the conventional sense, since the absence of a solid wall prevents boundary layer separation and, therefore, the pressure recovers in a short distance after the jet [100, 50]. The resulting momentum deficit is negligible in comparison to the wake behind a bluff body. Nevertheless, vortices of the classical von Kármán type are formed. As the Kelvin-Helmholtz instability, the von Kármán vortices are inherently irrotational and two-dimensional, interacting with turbulence as they develop.

3.4 Coherent structures and mixing

Eckart [25] first introduced the view of the turbulent mixing as a three-stage process of entrainment, stirring, and diffusion. Stirring is defined as the mechanical process of distributing fluids more uniformly in a given domain, increasing their interfacial area, while mixing is the diffusion process of substances across interfacial surfaces. Mixing is a molecular process, depending on diffusivities, while stirring is a purely kinematical process that depends on flow parameters.

Following this terminology, large-scale coherent structures promote entrainment in the flows they exist. The large-scale structures enhance the entrainment of fresh fluid in the turbulent region. In addition to that, the induced motions from the largest to the smallest eddies significantly enlarge the interfacial surface area between the fluid regions, stirring them. Despite their decisive contribution to the turbulent mixing, the coherent structures do not increase the molecular mixing directly. This fact is easily understood considering that it is possible to stir fluids that do not mix at all.

The turbulence, on the other hand, enhances the slow molecular diffusion. The effective diffusivity from a turbulent flow can be orders of magnitude higher than the molecular diffusivity, increasing the overall mixing rate.

Besides being different processes, entrainment, stirring and mixing are closely coupled in turbulent flows. The entrainment is responsible for bringing fresh fluid in the turbulent region, while stirring increases the interfacial area between them.

In the next chapter, the intermittency will be introduced to study the influence of entrainment and stirring in turbulent flows.

4 Intermittency

The previous chapters have presented two main features of turbulent free shear flows: turbulence and coherent structures. In the following chapter the concept of intermittency is introduced, a very powerful tool used to account for the influence of large-scale structures in turbulent flows.

4.1 Intermittency and intermittency function

All statistical quantities discussed in Chapter 2 are related to unconditional averages, *i.e.*, not considering intermittency. However, one of the characteristics of the flows in discussion in this work is that they have intermittent regions, meaning that in these regions fluid with different characteristics (for example turbulent and non-turbulent) can be found in an alternating matter.

Intermittency is present in all turbulent free shear flows [63]. All representations of flows that contain a mean "edge" between two fluid regions, like boundary layers or free shear flows, refer actually to the mean position of the interface between the two fluids. This phenomenon is illustrated in Figure 4.1, showing the time-resolved photograph of a liquid-phase turbulent jet where the flow is traced by means of fluorescent medium made visible using the laser-induced fluorescence technique. The surface that separates the two fluids is highly convoluted, with the typical large-scale structures being clearly visible. A fixed point toward the edge of the flow spends only a fraction of its time in the turbulent jet flow; the flow is called intermittent in this region.

Taking the turbulent jet from Figure 4.1 as an example, the quantitative description of the intermittency is possible using the intermittency function $I(t,x_i)$, which assumes value unity in the turbulent jet flow, and zero in the non-turbulent flow. Libby [63] arguments that it is possible to determine $I(x_i,t)$ experimentally from a careful interpretation of the output of one or more velocity signals, but the discrimination strategy required to do so is complex and subject to ambiguity because of velocity fluctuations within the non-turbulent fluid. An alternative definition of the intermittency function $I(x_i,t)$ is possible

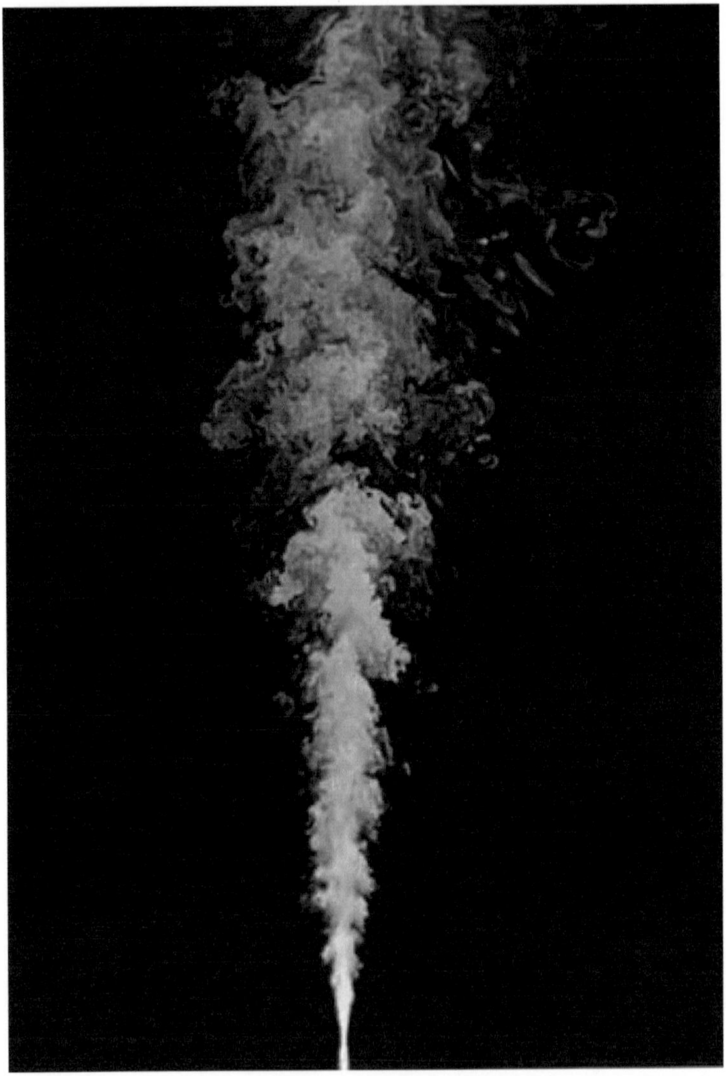

Figure 4.1: Liquid-phase turbulent jet symmetry plane slice. Gray scale codes jet-fluid mole fraction. Reynolds number approximately 10^4. Reproduced from [22].

in the case of the turbulent jet from Figure 4.1, as the turbulent flow can be associated with fluid having finite values of tracer concentration, since it is the jet flow that transports both momentum and tracer concentration in the quiescent fluid. Using this definition, and defining also a threshold value for the tracer selected to reflect the accuracy of the measurement or numerical setup, an intermittency function $I(x_i,t)$ can be constructed, which assumes value unity when the tracer value is higher than a small threshold value, and zero otherwise. This definition has also the advantage of being able to quantify the intermittency between two turbulent flows, as the jet in crossflow configuration in Chapter 7. Figure 4.2 shows the evolution of the axial velocity component and the mixture fraction at the outer border of a free jet, along with the intermittency function $I(x_i,t)$. The values were taken from the LES simulation described in Chapter 6, using the threshold value of 1%.

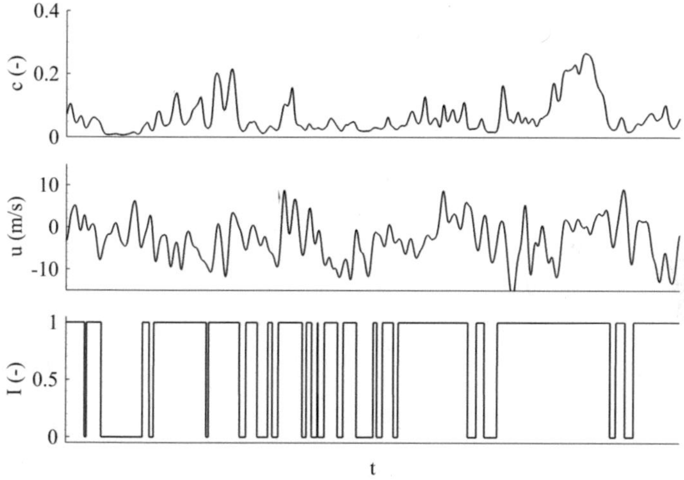

Figure 4.2: Evolution of passive scalar c, axial velocity component u and intermittency function I at the outer edge of a free jet.

Separate values for all statistical characteristics of a variable can be identified, simply by correlating it with the intermittency function. This process is called *conditional averaging*. Considering the variable c, the conditional average $\langle\ \rangle_t$ is defined as [10]

$$\langle c \rangle_t = C_t \equiv \frac{\overline{C \cdot I}}{\overline{I}} \equiv \lim_{t \to \infty} \frac{1}{\gamma T} \int_t^{t+T} c(x_i,t) I(x_i,t) dt. \tag{4.1}$$

where the subscript t represents the turbulent conditional average, and the intermittency factor $\gamma(x_i) \equiv \overline{I}$ represents the fraction of time a point is in the turbulent fluid.

Figure 4.3 shows a sketch of a jet flow, with radial profiles of the unconditional and conditional mean scalar and of the intermittency factor, variables C, C_t and γ, respectively. According to the intermittency factor γ, the flow is always turbulent at the jet axis, where it assumes the value unity, and never turbulent in the outer part of the flow, where it goes to zero. For all turbulent shear flows, experiments show that the profiles of γ are self-similar [1, 93]. The instantaneous interface between the two flows is called superlayer and is illustrated in Figure 4.3.

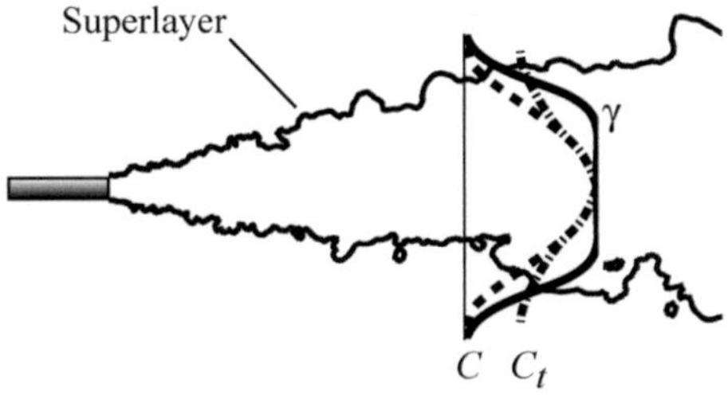

Figure 4.3: Sketch of a free jet flow, with the unconditional mean scalar C, the conditional mean C_t and the intermittency factor γ, from the experimental data of [2], and the instantaneous position of the superlayer from a snapshot of the LES shown in Chapter 6.

Experimental data of Antonia *et al.* [2] for a turbulent heated jet in a co-flowing stream can be seen in Figures 4.4 to 4.6. The radial distance r was normalized with its half-radius value, resulting in the variable $r/r_{1/2}$, while the axial velocity component U and the temperature T were normalized with their centerline values, resulting in the variables U/U_c and T/T_c. Using the intermittency function, the conditional averages in the turbulent, jet flow (subscript t) and in the non-turbulent, ambient fluid flow (subscript n) were calculated.

Figure 4.4 shows the radial variation of unconditioned mean axial velocity U/U_c and temperature T/T_c, along with the intermittency factor γ. It should be noted that the profiles are not symmetric as they should be, indicating some level of inaccuracy in the measurements. Nevertheless, while the velocity and temperature profiles are approximately similar, with a peak value at the axis that steadily decreases toward their edge values, γ is almost constant from the center of the jet until the jet half-radius ($r/r_{1/2}$=1), decreasing rapidly towards zero at the jet edge. The region of influence of the intermittency is restricted to this outer region.

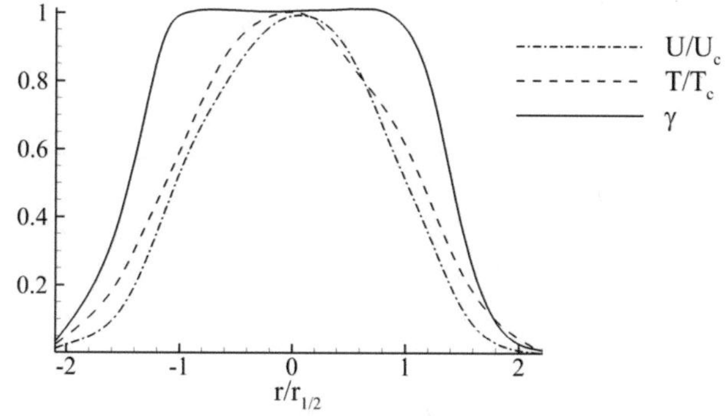

Figure 4.4: Normalized mean axial velocity U/U_c, temperature T/T_c and intermittency function γ across the jet, from the experimental data of [2].

The radial variations of the unconditioned mean temperature T/T_c, the mean temperature conditioned to the turbulent jet flow T_t/T_c and to the non-turbulent ambient fluid T_n/T_c and the intermittency factor γ are shown in Figure 4.5. The positive value of T_n/T_c towards the jet center, approaching 0.1 at $r/r_{1/2} = 0.6$, is certainly related to some inaccuracy in the measurements. The value of T_n/T_c should be zero by definition, as the ambient fluid maintains its temperature even after being entrained by the jet, otherwise it should be considered jet fluid.

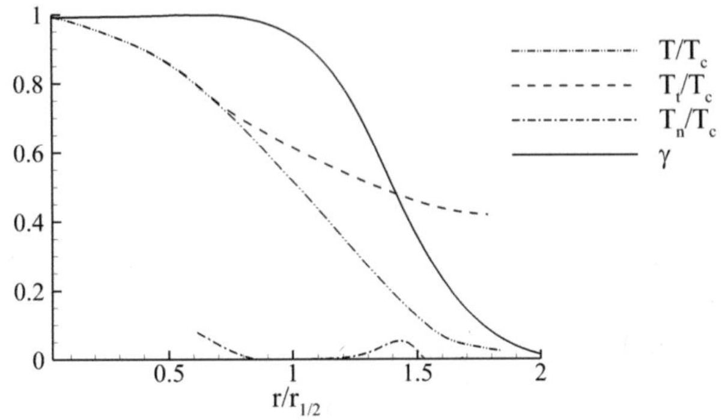

Figure 4.5: Unconditional mean temperature T/T_c, the conditioned turbulent T_t/T_c and non-turbulent T_n/T_c mean temperature and the intermittency factor γ, from the experimental data of [2].

The value of T_t/T_c is considerably higher than the local mean temperature in the outer region. This result indicates that the diminishing value of the unconditional mean scalar when approaching the jet edge is a consequence of the diminishing fraction of time that fluid with an almost constant value of scalar spends in this region. This fact has great importance for mixing-sensitive systems. The conditioned mean temperature tends to a value of about 0.4 in the outer, intermittent region of the jet, while the unconditioned mean temperature tends to zero. The conditioned mean axial velocity in Figure 4.6, on the other hand, tends to zero when the outer region is approached, as does the unconditioned mean velocity. It is evident that the difference between conditioned and unconditioned values is considerably higher in the temperature field (Figure 4.5) than in the corresponding axial velocity field (Figure 4.6).

The different behavior of velocity and passive scalar fields when confronted to intermittent flows is the key to understand the otherwise paradoxical results of the steady-state simulations of the jet in crossflow, in which velocity fields showed good agreement with the measurements while the passive scalar field presented significant deviations. These results will be discussed in more detail in Chapters 6 and 7.

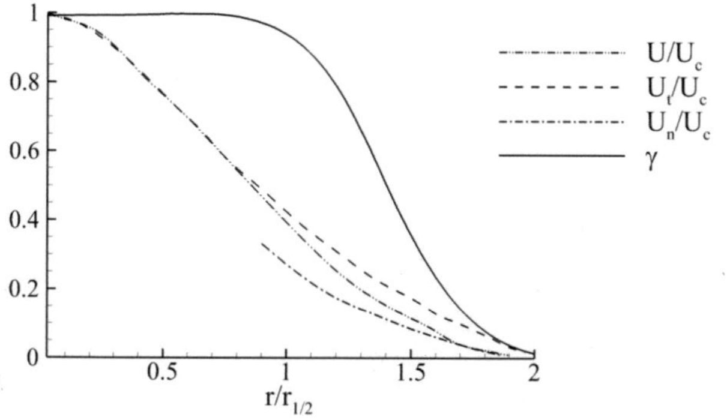

Figure 4.6: Unconditional mean axial velocity U/U_c, conditional mean axial velocity U_t/U_c and ambient fluid mean axial velocity U_n/U_c, from the experimental data of [2].

5 Simulation Theory

In Chapters 2 to 4, the fundamental physics of turbulent flows with and without reaction have been presented.

This chapter introduces the numerical tools used in the simulations. It begins with an overview of turbulence modeling and simulation, followed by the models used in Reynolds-averaged context and in large eddy simulations. The chapter proceeds describing the reaction models used in the simulation of combustion systems, along with a brief introduction about the presumed JPDF model. The final section highlights the details of the flow solver used in the simulations.

5.1 Turbulence modeling and simulation

In order to study turbulent flows numerically, the turbulence can be *simulated* or *modeled*. In the *simulation* of turbulent flows, a time-dependent set of equations is solved, representing an approximation of one realization of the turbulent flow. In contrast, in a *turbulence model*, equations for mean quantities, for example U, k and ε, are solved. The influence of the turbulence on the mean quantities is modeled.

Examples of turbulence simulation are direct numerical simulation (DNS) and large eddy simulation (LES). In DNS, one realization of the flow is fully simulated, including all length and time scales. These requirements make DNS extremely computationally intensive and inappropriate for the high turbulence level of the flows studied in this work. Annex A.1 shows an example of the prohibitive computational resources required for a DNS. In LES, equations of the filtered velocity field are solved, representing the large-scale turbulent flow. The influence of the small-scale motions is modeled, making the whole simulation much more affordable than DNS, but the stochastic nature of turbulent solutions of the Navier-Stokes equations is retained.

In the case of Reynolds-averaged Navier-Stokes (RANS), the whole turbulence is modeled. In the two-equation models used in this work, the Reynolds stresses, which describe the influence of the turbulence on the mean velocity, are obtained using a turbulent viscosity approach. RANS turbulence models can be used in both steady-state and unsteady simulations. Despite the time dependence and the ability of solving large-scale structures, unsteady RANS (URANS) does not simulate the turbulence directly, only modeling its statistics.

In flows that are statistically stationary, the solution of the RANS equations does not depend on time, and both steady-state and unsteady simulations lead to the same results. In contrast, for statistically non-stationary flows, the mean values fluctuate with time, and only the unsteady simulation is appropriate. The turbulence remains fully modeled; the velocity fluctuations are associated with large-scale coherent structures that are not turbulent in nature. These simulations are generally called unsteady RANS or URANS simulations.

Various hybrid approaches have been proposed, combining RANS and LES: Travin *et al.* [113] proposed the detached eddy simulation (DES), and Menter and Egorov [74] proposed the scale-adaptive simulation (SAS). The two models work limiting the value of the turbulent viscosity in some areas of the flow (for example away from walls), promoting the development of large-scale structures there. They should not be confused with URANS, as the turbulence is not fully modeled in this case. The main criticism raised against the hybrid approaches is that in the process of limiting the turbulent viscosity, a fraction of the turbulent kinetic energy gets lost. In the RANS part of the flow, there are no velocity fluctuations due to turbulence, as they are fully modeled. In the transition to the LES part of the flow, the turbulent viscosity is numerically reduced, however without the introduction of the corresponding velocity fluctuations. In some applications this inconsistency is negligible. In the case of the jet in crossflow, however, this inconsistency had an adverse effect on the transition of the jet flow issuing from the pipe, leading to erroneous prediction of the jet penetration depth.

In face of the impossibility of using DNS and the issues of DES/SAS simulations, three simulation strategies have been used in this work: steady-state and unsteady RANS simulations, described in Sections 5.2 and 5.3, and LES, described in Section 5.4.

5.2 RANS turbulence models

This section presents the RANS turbulence models for incompressible flows. The differences introduced by compressible flows are detailed in the next section.

The averaging procedure of the Reynolds-averaged Navier-Stokes (RANS) equations (see Section 2.4.1) introduces additional unknowns so that the resulting system of equations is not closed. Various methods can be used to overcome this difficulty, the so-called eddy viscosity models being the most widespread of them.

The eddy viscosity ν_t was first postulated by Prandtl [94]. In direct analogy to the laminar viscosity, it is the product of a turbulent velocity scale u_t and a turbulent length scale ℓ_t

$$\nu_t = \text{constant} \cdot u_t \cdot \ell_t. \tag{5.1}$$

As the basis of the turbulent velocity scale, Prandtl chose the kinetic energy (per unit mass) of the turbulent fluctuations

$$k = \frac{1}{2}\overline{u_i' u_i'} = \frac{1}{2}\left(\overline{u'u'} + \overline{v'v'} + \overline{w'w'}\right). \tag{5.2}$$

Using dimensional arguments, the characteristic velocity scale is defined as

$$u_t = \sqrt{\frac{2}{3}k}. \tag{5.3}$$

The Boussinesq approximation relates the Reynolds stress tensor to the mean strain rate tensor, employing the eddy viscosity [121]

$$-\overline{u_i' u_j'} = 2\nu_t S_{ij} - \frac{2}{3}k\delta_{ij} \tag{5.4}$$

where δ_{ij} is the Kronecker delta and S_{ij} is the mean strain rate tensor, defined as

$$S_{ij} = \frac{1}{2}\left(\frac{\partial U_i}{\partial x_j} + \frac{\partial U_j}{\partial x_i}\right). \tag{5.5}$$

A transport equation for the turbulent kinetic energy k can be derived multiplying the instantaneous Navier-Stokes equations (2.1 and 2.2) by the fluctuating velocity u'_i, averaging the resulting equations and doing some algebra exercises

$$\frac{\partial k}{\partial t}+U_j\frac{\partial k}{\partial x_j}=-\overline{u'_iu'_j}\frac{\partial U_i}{\partial x_j}-\varepsilon+\frac{\partial}{\partial x_j}\left[\nu\frac{\partial k}{\partial x_j}-\frac{1}{2}\overline{u'_iu'_iu'_j}-\frac{1}{\rho}\overline{p'u'_j}\right] \quad (5.6)$$

where ε, the dissipation rate per unit mass, is defined as

$$\varepsilon=\nu\overline{\frac{\partial u'_i}{\partial x_k}\frac{\partial u'_i}{\partial x_k}}. \quad (5.7)$$

Following the analysis of Wilcox [121], the various terms in equation (5.6) represent different physical processes. The sum of the *unsteady term* and the *convection*, both on the left-hand side of the equation, represent the substantial derivative of k. The first term on the right-hand side represents the rate at which kinetic energy is transferred from the mean flow to the turbulence, and it is known as *production*. The next term is the *dissipation*, and represents the rate at which turbulent kinetic energy is converted into internal energy. The first term in the square brackets is the *molecular diffusion* of turbulent kinetic energy. The triple correlation of velocity fluctuations is known as *turbulent transport* and represents the rate at which the turbulence energy is transported by turbulent fluctuations. The last term is called *pressure diffusion* and represents the turbulent transport via the correlation of pressure and velocity fluctuations.

Production, dissipation, turbulent transport and pressure diffusion add unknown correlations and have to be modeled, while the unsteady term, convection and molecular diffusion can be solved directly. To close the equation the Reynolds stress tensor, turbulent transport, pressure diffusion and dissipation have to be modeled.

The Reynolds stress tensor is modeled using the Boussinesq approximation, Equation (5.4), closing the production term. The turbulent transport and the pressure diffusion are generally modeled together, using a gradient-diffusion approach

$$\frac{1}{2}\overline{u'_iu'_iu'_j}+\frac{1}{\rho}\overline{p'u'_j}=\frac{\nu_t}{\sigma_k}\frac{\partial k}{\partial x_j}, \quad (5.8)$$

where σ_k is the turbulent Prandtl number for k.

The manner in which the dissipation ε is modeled is not unique amongst the different turbulence models. It is sufficient at this point to note that it is proportional to the turbulent length scale

$$\varepsilon \approx k^{3/2}/\ell_t. \tag{5.9}$$

Hence, the turbulent length scale has to be prescribed in order to close the system. The different approaches used to model the turbulent length scale lead to different turbulence models.

Combining equations (5.6) and (5.8) results in the modeled version of the turbulent kinetic energy equation

$$\frac{\partial k}{\partial t} + U_j \frac{\partial k}{\partial x_j} = -\overline{u_i' u_j'} \frac{\partial U_i}{\partial x_j} - \varepsilon + \frac{\partial}{\partial x_j}\left[\left(\nu + \frac{\nu_t}{\sigma_k}\right)\frac{\partial k}{\partial x_j}\right] \tag{5.10}$$

5.2.1 The k-ε model

The standard k-ε model [47, 59] was by far the most popular two-equation model until the last decade of the twentieth century. In the derivation of the model, the assumption is that the flow is fully turbulent and that the effects of molecular viscosity are negligible. The standard k-ε model is therefore valid only for fully turbulent, high Reynolds number flows.

The turbulent kinetic energy k is solved using equation (5.10), while its dissipation rate, ε, is solved using the following transport equation:

$$\frac{\partial \varepsilon}{\partial t} + U_i \frac{\partial \varepsilon}{\partial x_i} = \frac{\partial}{\partial x_j}\left[\left(\nu + \frac{\nu_t}{\sigma_\varepsilon}\right)\frac{\partial \varepsilon}{\partial x_j}\right] - C_{1\varepsilon}\frac{\varepsilon}{k}\overline{u_i' u_j'}\frac{\partial U_i}{\partial x_j} - C_{2\varepsilon}\frac{\varepsilon^2}{k} \tag{5.11}$$

where $C_{1\varepsilon}$ and $C_{2\varepsilon}$ are constants and σ_ε is the turbulent Prandtl number for ε. The various terms in equation (5.11) represent, from left to right: *Unsteady Term*, *Convection*, *Diffusion*, *Production* and *Destruction*.

To evaluate the turbulent production in a manner consistent with the Boussinesq hypothesis, it is calculated as

$$-\overline{u_i' u_j'}\frac{\partial U_i}{\partial x_j} = \nu_t S^2 \tag{5.12}$$

where S is the modulus of the mean strain rate tensor, defined as

$$S \equiv \sqrt{2S_{ij}S_{ij}}. \tag{5.13}$$

The turbulent length scale is defined as

$$\ell_t = C_\mu \frac{k^{3/2}}{\varepsilon}, \tag{5.14}$$

resulting in the turbulent viscosity ν_t, introduced in Equation (5.1)

$$\nu_t = C_\mu \frac{k^2}{\varepsilon} \tag{5.15}$$

where C_μ is a constant.

The most used values for the constants $C_{1\varepsilon}, C_{2\varepsilon}, C_\mu, \sigma_k$, and σ_ε are [59]

$$C_{1\varepsilon} = 1.44, \;\; C_{2\varepsilon} = 1.92, \;\; C_\mu = 0.09, \;\; \sigma_k = 1.0, \;\; \sigma_\varepsilon = 1.3$$

5.2.2 The Shear-Stress-Transport (SST) model

The length scale needed to define the eddy viscosity can be derived not only from the dissipation rate ε, but also from other quantities. The k-ω model is an empirical model based on model transport equations for the turbulent kinetic energy k and the dissipation per unit turbulent kinetic energy ω, which can also be thought of as the ratio of ε to k [121]. The reciprocal of ω is the time scale on which dissipation of turbulent energy occurs. The major advantage over the standard k-ε model is that it can be integrated to the wall without the need of damping functions. However the k-ω model is highly sensitive to the free stream boundary conditions, which limits its application.

The shear-stress transport (SST) k-ω model was developed by Menter [73]. The objective was to unite the advantages of two models: the robustness and accuracy of the k-ω model in the near-wall region and the free-stream independence of the k-ε model in the far field. The SST model was developed converting the k-ε model into a k-ω formulation and multiplying both k-ω and k-ε by a blending function. The blending function is designed to be one in the near-wall region, activating the k-ω model, and zero in the free-stream, activating the transformed k-ε model. The definition of the turbulent viscosity is also modified to account for the transport of the turbulent shear stress.

Due to the enhancements introduced by Menter [73], the SST k-ω model is potentially more accurate and reliable than the standard k-ω model. The high-Reynolds version of the model was used in this work.

The k equation is

$$\frac{\partial k}{\partial t}+U_i\frac{\partial k}{\partial x_i}=\frac{\partial}{\partial x_j}\left[\left(\nu+\frac{\nu_t}{\sigma_k}\right)\frac{\partial k}{\partial x_j}\right]+G_k-\beta^*k\omega \tag{5.16}$$

and the ω equation is

$$\frac{\partial \omega}{\partial t}+U_i\frac{\partial}{\partial x_i}=\frac{\partial}{\partial x_j}\left[\left(\nu+\frac{\nu_t}{\sigma_\omega}\right)\frac{\partial \omega}{\partial x_j}\right]+G_\omega-\beta\omega^2+D_\omega. \tag{5.17}$$

In these equations, G_k represents the production of turbulent kinetic energy due to mean velocity gradients

$$G_k=\min(-\overline{u_i'u_j'}\frac{\partial U_i}{\partial x_j},10\cdot\beta^*k\omega), \tag{5.18}$$

and G_ω represents the production of ω

$$G_\omega=-\rho\overline{u_i'u_j'}\frac{\partial U_i}{\partial x_j}\frac{\alpha}{\nu_t}. \tag{5.19}$$

D_ω represents the cross-diffusion term

$$D_\omega=2\left(1-F_1\right)\sigma_{\omega,2}\frac{1}{\omega}\frac{\partial k}{\partial x_j}\frac{\partial \omega}{\partial x_j}, \tag{5.20}$$

where σ_k and σ_ω are the turbulent Prandtl numbers for k and ω, given by

$$\sigma_k=\frac{1}{F_1/\sigma_{k,1}+(1-F_1)/\sigma_{k,2}} \tag{5.21}$$

$$\sigma_\omega=\frac{1}{F_1/\sigma_{\omega,1}+(1-F_1)/\sigma_{\omega,2}}. \tag{5.22}$$

The turbulent viscosity ν_t is computed as

$$\nu_t = \frac{k}{\omega}\frac{1}{\max\left[\frac{1}{\alpha^*}, \frac{SF_2}{a_1\omega}\right]}, \tag{5.23}$$

where S is the strain rate magnitude and α^* is a constant. The blending functions, F_1 and F_2, are given by

$$F_1 = \tanh\left(\Phi_1^4\right) \tag{5.24}$$

$$\Phi_1 = \min\left[\max\left(\frac{\sqrt{k}}{0.09\omega y}, \frac{500\nu}{y^2\omega}\right), \frac{4k}{\sigma_{\omega,2}D_\omega^+ y^2}\right] \tag{5.25}$$

$$D_\omega^+ = \max\left[2\frac{1}{\sigma_{\omega,2}}\frac{1}{\omega}\frac{\partial k}{\partial x_j}\frac{\partial \omega}{\partial x_j}, 10^{-10}\right] \tag{5.26}$$

$$F_2 = \tanh\left(\Phi_2^2\right) \tag{5.27}$$

$$\Phi_2 = \max\left[2\frac{\sqrt{k}}{0.09\omega y}, \frac{500\mu}{\rho y^2\omega}\right] \tag{5.28}$$

where y is the distance to the next surface and D_ω^+ is the positive portion of the cross-diffusion term (Equation 5.20).

The model constants are:

$$\sigma_{k,1} = 1.176,\ \ \sigma_{k,2} = 1.0,\ \ \sigma_{\omega,1} = 2.0,\ \ \sigma_{\omega,2} = 1.168$$
$$a_1 = 0.31,\ \ \beta_{i,1} = 0.075,\ \ \beta_{i,2} = 0.0828,\ \ \alpha^* = 1$$

5.2.3 Turbulent mixing model

The Reynolds-averaged equation of a passive scalar, Equation (2.19), also needs to be closed.

By analogy to the Boussinesq approximation (Equation 5.4), the turbulent transport of a scalar is assumed to be proportional to the mean scalar gradient

$$-\overline{u_i'c'} = D_t\frac{\partial C}{\partial x_i} \tag{5.29}$$

where D_t is the scalar turbulent diffusivity.

One option to evaluate D_t is to consider that the mechanisms of turbulent transport of a passive scalar and of momentum are essentially the same, *i.e.*, enhanced mixing due to the turbulent motions. The value of D_t is expected to be close to ν_t. The ratio between ν_t and D_t leads to the definition of the turbulent Schmidt or Prandtl number

$$\sigma_t = \frac{\nu_t}{D_t} \tag{5.30}$$

which is constant and equal to 0.9, when not otherwise noted.

5.2.4 Wall treatment

The physics of turbulence in the vicinity of walls is considerably different from free shear flows. It is therefore necessary to use appropriate turbulence models in the near-wall region. For the most general and detailed treatment, low-Reynolds versions of the turbulence models should be used, which are able to predict the flow down to the viscous sublayer. However, in order to resolve correctly all the near-wall details, the computational grid needs to be extremely fine in this region, which increases the computational cost of the whole simulation.

It is possible to approximate the effects of the wall without explicitly resolving the near-wall region using the so called wall-functions. Wall-functions are a simplified model of turbulence, which predict the near-wall profiles of the velocity, k and ε (or ω) variables. It is used to bridge the regions of high gradients near the wall with the high-Re turbulence model in the rest of the domain.

The law of the wall for mean velocity yields [121]

$$U^+ = \frac{U}{u_\tau} = \frac{1}{\kappa} \ln(Ey^+) \tag{5.31}$$

where the friction velocity u_τ and the dimensionless wall distance y^+ are defined as

$$u_\tau = \sqrt{\frac{\tau_W}{\rho}}; \; y^+ = \frac{u_\tau y_P}{\nu}, \tag{5.32}$$

where τ_W is the wall shear stress, y_P is the distance from point P to the wall, ν is the kinematic viscosity of the fluid, $\kappa = 0.4187$ is the von Kármán constant and E=9 is an empirical constant.

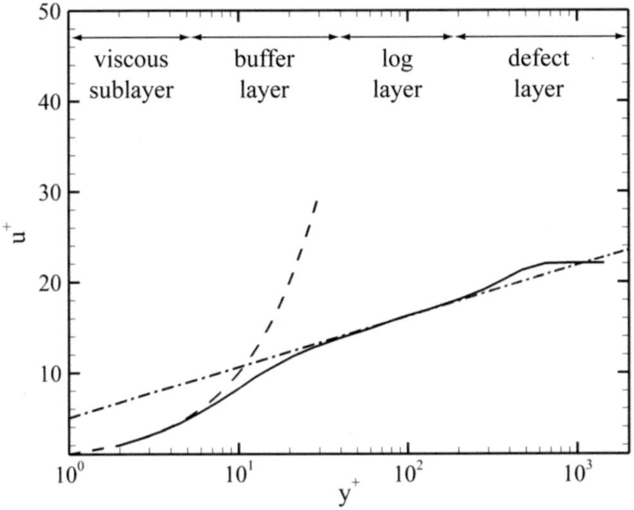

Figure 5.1: Mean velocity profile of a boundary layer in wall units. Dashed line, Equation (5.33); dash-dot line, Equation (5.31); solid line, boundary layer DNS of Spalart [107].

The logarithmic law is known to be valid for y^+ values higher than 30. In the region where y^+ is smaller than 5, called viscous sublayer, the linear relationship is valid

$$U^+ = y^+. \tag{5.33}$$

The region between $5 < y^+ < 30$ is known as buffer layer. Figure 5.1 shows the viscous sublayer, the buffer layer and the log layer, along with the boundary layer DNS of Spalart [107] for Re_θ = 1441 (Re_θ is the Reynolds number defined with the momentum thickness θ). The logarithmic and the linear profiles intersect at y^+=11.63, however placing the first point of the grid in the buffer region $5 < y^+ < 30$ should be avoided, as the deviation between the logarithmic and linear profiles and the correct value of U^+ becomes large.

Although not clearly defined, the form of the law of the wall in the buffer layer can be approximated. In Section 5.4.3, the Spalding's law of the wall [108] is presented, which is a fit of the viscous, buffer and logarithmic regions in one equation and is used in the LES.

There are different methods of implementing the law of the wall. For the RANS simulations, an implementation based on the proposal of Launder and Spalding [59] has been used, in which the k equation is solved in the whole domain including the wall-adjacent cells. The boundary condition for k at the walls is

$$\frac{\partial k}{\partial n} = 0 \tag{5.34}$$

where n is the local coordinate normal to the wall.

The production of turbulent kinetic energy k and its dissipation rate ε at the cells at the walls, which are the source terms in the k equation, are computed assuming the local equilibrium hypothesis, *i.e.*, that the production of k and ε balance in the control volumes at the walls. The production of k is defined as

$$\overline{u_i' u_j'} \frac{\partial U_i}{\partial x_j} \approx \tau_w \frac{\partial U}{\partial y} = \tau_w \frac{u_\tau}{\kappa y_P} \tag{5.35}$$

using Equation (5.31) to calculate the derivative. Approximating u_τ with the relation $C_\mu^{1/4} k^{1/2}$, as proposed by Launder and Spalding [59], yields the production of k as

$$\overline{u_i' u_j'} \frac{\partial U_i}{\partial x_j} \approx \tau_w \frac{C_\mu^{1/4} k^{1/2}}{\kappa y_P}, \tag{5.36}$$

which is the function implemented in the CFD code. The dissipation rate ε is computed as

$$\varepsilon_P = \frac{C_\mu^{3/4} k_P^{3/2}}{\kappa y_P}, \tag{5.37}$$

where the subscript P represents the value of the variable at the point P. The ε equation is not solved at the wall-adjacent cells, but instead the value is deduced applying equation (5.37).

In this work, all scalars were computed under the assumption of impermeable walls. In this case the boundary condition at the wall is

$$\frac{\partial C}{\partial n} = 0, \tag{5.38}$$

here n is the local coordinate normal to the wall.

5.2.5 Steady state and unsteady RANS simulation

It is important to note that, as already expressed in Section 2.4.1, being the most general Reynolds average, the ensemble average was used to derive all averaged equations so far.

In the case of a turbulent flow that is statistically stationary, the ensemble average commutes to the time average, and the time derivative of all transport equation are equal to zero. This fact has implications for the numerical solution of the equation system, as the time dependence can be dropped. Such simulations are generally called *steady state* RANS or only RANS simulations.

In the case of statistically unsteady flow, the transport equations retain their time derivatives. The numerical solution in this case iterates in time. These simulations are generally called *unsteady* RANS or URANS simulations.

5.3 RANS models for compressible flows

By definition, a compressible flow is one in which significant density changes occur, even when pressure changes are small [121]. It includes low speed flows with large heat transfer rates, as the reacting flows in this work.

The equations for compressible flows, derived using Favre averaging (see Section 2.5.2), are formally identical to the classical Reynolds-averaged equations for incompressible flows (see Section 2.4.1). Favre averaging simplifies the transport equations, avoiding the need to model most of the correlations involving density fluctuations. Some correlations, however, still require modeling.

Differently from the definition used for incompressible flows, the viscous stress tensor for a compressible flow is

$$t_{ij} = 2\mu s_{ij} - \frac{2}{3}\mu \frac{\partial u_k}{\partial x_k}\delta_{ij}, \tag{5.39}$$

where s_{ij} is the instantaneous strain rate tensor and δ_{ij} is the Kronecker delta. The difference lies in the last term, which is zero in incompressible flows due to continuity. The transport equation of the turbulent kinetic energy k for compressible flows is [121]

$$\begin{aligned} \overline{\rho}\frac{\partial k}{\partial t} + \overline{\rho}\tilde{u}_j\frac{\partial k}{\partial x_j} &= -\overline{\rho}\widetilde{u_i''u_j''}\frac{\partial \tilde{u}_i}{\partial x_j} + \frac{\partial}{\partial x_j}\left[\overline{t_{ij}u_i''} - \frac{1}{2}\overline{\rho u_i''u_i''u_j''} - \overline{p'u_j''}\right] \\ &\quad -\overline{\rho}\varepsilon - \overline{u_i''}\frac{\partial P}{\partial x_i} + \overline{p'\frac{\partial u_i''}{\partial x_i}} \end{aligned} \tag{5.40}$$

where $\overline{\rho}\varepsilon$, the Favre-averaged dissipation rate, is defined as

$$\overline{\rho}\varepsilon = \overline{\frac{1}{2}t_{ij}\frac{\partial u_i''}{\partial x_k}\frac{\partial u_i''}{\partial x_k}}. \tag{5.41}$$

Equation (5.40) is very similar to the equation for incompressible flows, Equation (5.6), except the last two terms: the *pressure work* $\overline{u_i''}\partial P/\partial x_i$ and *pressure dilatation* $\overline{p'\partial u_i''/\partial x_i}$ terms. In incompressible flows, the pressure work term vanishes because the time average of u_i'' is zero when density fluctuations are zero, and the pressure dilatation vanishes because the fluctuating field has zero divergence.

5.3.1 The compressible k-ε model

The RANS simulations of compressible flows in this work have used the standard k-ε model. Efforts can be found in the literature to close the pressure work and pressure dilatation terms of the k equation, as shown in the textbook of Wilcox [121]. He states, however, that all these terms appear to be negligible for the Reynolds numbers of practical interest. The modeled equation for k turns to be very similar to the equation for incompressible flows

$$\overline{\rho}\frac{\partial k}{\partial t}+\overline{\rho}\tilde{u}_j\frac{\partial k}{\partial x_j}=-\overline{\rho}\widetilde{u_i''u_j''}\frac{\partial \tilde{u}_i}{\partial x_j}-\overline{\rho}\varepsilon+\frac{\partial}{\partial x_j}\left[\left(\mu+\frac{\mu_t}{\sigma_k}\right)\frac{\partial k}{\partial x_j}\right] \tag{5.42}$$

where σ_k is the turbulent Schmidt number for k.

The equation for the dissipation rate $\overline{\rho}\varepsilon$ is also similar to its counterpart for incompressible flows

$$\overline{\rho}\frac{\partial \varepsilon}{\partial t}+\overline{\rho}\tilde{u}_j\frac{\partial \varepsilon}{\partial x_i}=\frac{\partial}{\partial x_j}\left[\left(\mu+\frac{\mu_t}{\sigma_\varepsilon}\right)\frac{\partial \varepsilon}{\partial x_j}\right]-C_{1\varepsilon}\frac{\varepsilon}{k}\overline{\rho u_i'u_j'}\frac{\partial \tilde{u}_i}{\partial x_j}$$
$$-C_{2\varepsilon}\overline{\rho}\frac{\varepsilon^2}{k} \tag{5.43}$$

with the same constants as the incompressible model shown in Section 5.2. The turbulent viscosity μ_t is defined as

$$\mu_t=\overline{\rho}C_\mu\frac{k^2}{\varepsilon} \tag{5.44}$$

5.3.2 Reynolds stress and fluxes

The unclosed terms in the Favre-averaged equations (2.32) to (2.35) are modeled in the following way

Reynolds stress tensor

As for incompressible flows, the Boussinesq approximation is used, with μ_t denoting the turbulent viscosity

$$-\overline{\rho}\widetilde{u_i''u_j''}=2\mu_t S_{ij}-\frac{1}{3}\frac{\partial \tilde{u}_k}{\partial x_k}\delta_{ij}-\frac{2}{3}\overline{\rho}k\delta_{ij} \tag{5.45}$$

Reynolds flux vector

By analogy to the Boussinesq approximation, the Reynolds flux vector for mass transfer is modeled as

$$-\bar{\rho}\widetilde{u_i''Y_c''} = \frac{\mu_t}{\sigma_t}\frac{\partial \tilde{c}}{\partial x_i} \tag{5.46}$$

and for heat transfer as

$$-\bar{\rho}\widetilde{u_i''h''} = \frac{\mu_t}{\sigma_t}\frac{\partial \tilde{h}}{\partial x_i} \tag{5.47}$$

where σ_t is the turbulent Schmidt/Prandtl number. As in the case of the incompressible model, σ_t is constant and equal to 0.9, when not otherwise noted.

5.4 Large Eddy Simulation

Large eddy simulation (LES) is a popular technique for simulating turbulent flows. It was first proposed by Kolmogorov [54], using the fact that large eddies of the flow are geometry-dependent while the smaller scales more universal. The LES approach is a compromise between DNS and RANS; direct simulation being applied to the large-scales, while the small scales have their effect modeled. The idea behind LES is that the large eddies contain most of the energy and do most of the transport of conserved properties. The large eddies are more dependent on the geometry and boundary conditions, while small eddies tend to be more isotropic, independent of the geometry and are consequently more universal. The modeling is thus simplified when only the small eddies are considered, and the errors introduced by their modeling should be small.

Filtering is essentially a mathematical manipulation of the exact Navier-Stokes equations, in which the eddies smaller than the filter size are filtered out. The filter size is usually the mesh size when spatial filtering is employed. Like the Reynolds averaging introduced in Chapter 2, the filtering process creates additional unknown terms that must be modeled. The statistics of the time varying flow field such as time averages and root-mean-square values can be calculated during the time-dependent simulation.

The LES models for incompressible flows are derived in detail in the following section. Section 5.4.4 shows the extension to compressible flows.

5.4.1 Volume averaging

Volume averaging corresponds to a filtering operation, with the scales larger than the filter being solved directly and the smaller scales being averaged. Leonard [61] defines a generalized filter as a convolution integral

$$\overline{u_i}(x_i,t) = \int G(x_i - x_i')u_i(x_i',t)dx_i', \tag{5.48}$$

where $\overline{u_i}$ denotes the filtered velocity, and the filter function G satisfies the normalization condition

$$\int G(x_i - x_i')dx_i' = 1. \tag{5.49}$$

Defining the residual field by

$$u_i'(x_i,t) \equiv u_i(x_i,t) - \overline{u_i}(x_i,t) \tag{5.50}$$

results in a decomposition very similar to the Reynolds decomposition

$$u_i(x_i,t) = \overline{u_i}(x_i,t) + u_i'(x_i,t). \tag{5.51}$$

One important difference is that the filtered residual is not zero in general

$$\overline{u_i'}(x_i,t) \neq 0. \tag{5.52}$$

The overbar represents the filtering operation in the above equations.

The transport equations employed for LES are derived by filtering the instantaneous Navier-Stokes equations, Equations 2.1 and 2.2. The filtering process effectively filters out eddies whose scales are smaller than the filter width. The resulting equations thus govern the dynamics of large eddies. Filtering the Navier-Stokes equations, one obtains for the continuity

$$\frac{\partial \overline{u_i}}{\partial x_i} = 0 \tag{5.53}$$

and for the momentum

$$\frac{\partial \overline{u_i}}{\partial t} + \frac{\partial \overline{u_i u_j}}{\partial x_j} = -\frac{1}{\rho}\frac{\partial \overline{p}}{\partial x_i} + \frac{\partial}{\partial x_j}\left(\nu \frac{\partial \overline{u_i}}{\partial x_j}\right). \tag{5.54}$$

Since the convective flux $\overline{u_i u_j}$ is not equal to $\overline{u}_i \overline{u}_j$, a modeling approximation has to be introduced

$$\overline{u_i u_j} = \overline{u}_i \overline{u}_j + \tau_{ij} \tag{5.55}$$

where the subgrid scale stress tensor τ_{ij} is defined as

$$\tau_{ij} = \underbrace{\overline{\overline{u}_i \overline{u}_j} - \overline{u}_i \overline{u}_j}_{L_{ij}} + \underbrace{\overline{u'_i \overline{u}_j} + \overline{\overline{u}_i u'_j}}_{C_{ij}} + \underbrace{\overline{u'_i u'_j}}_{R_{ij}}. \tag{5.56}$$

The tensors L_{ij}, C_{ij} and R_{ij} are known as the Leonard stress, cross-term stress and the subgrid scale (SGS) stress, respectively. While the SGS stress as a whole is Galilean invariant (independent of inertial frame), the cross-term and Leonard stresses are not. It follows that the correlations used to model these stresses are approximations, which contain errors that cannot be eliminated. For these reasons, in the modern literature all terms are modeled together, instead of decomposing the SGS stresses [20].

The conservation equation for a passive scalar can be derived in a similar way resulting in

$$\frac{\partial \overline{c}}{\partial t} + \frac{\partial \overline{u_j c}}{\partial x_j} = \frac{\partial}{\partial x_j}\left(D \frac{\partial \overline{c}}{\partial x_j}\right). \tag{5.57}$$

The convective flux $\overline{u_j c}$ has also to be modeled

$$\overline{u_j c} = \overline{u}_j \overline{c} + q_j, \tag{5.58}$$

where the subgrid scale flux vector q_j is modeled as a whole, in the same fashion as the τ_{ij} in the momentum equation.

As in the Reynolds-averaged equations, the subgrid contributions cannot be represented directly by the filtered velocity and scalar fields and have to be modeled. The subgrid scale models used to close the equations are discussed in the following section.

5.4.2 SGS modeling

The first model for the subgrid scale stresses was developed by Smagorinsky [104]. The model employs the Boussinesq approximation (Equation 5.4) as in the RANS models, which assumes that the SGS stresses τ_{ij} follow a gradient-diffusion process. Consequently, τ_{ij} is given by

$$\tau_{ij} - \frac{1}{3}\tau_{kk}\delta_{ij} = -2\nu_t S_{ij} \tag{5.59}$$

where ν_t is the subgrid scale turbulent viscosity. Only the anisotropic part of the subgrid scale stresses τ_{ij} are resolved by the subgrid scale model; the isotropic part is modeled together with the static pressure term. S_{ij} is the strain rate tensor for the resolved scale defined by

$$S_{ij} \equiv \frac{1}{2}\left(\frac{\partial \overline{u}_i}{\partial x_j} + \frac{\partial \overline{u}_j}{\partial x_i}\right). \tag{5.60}$$

The subgrid scale turbulent flux of a passive scalar c (see Equation (5.58)) is modeled using a subgrid scale turbulent Schmidt number σ_t

$$q_j = -\frac{\nu_t}{\sigma_t}\frac{\partial \overline{c}}{\partial x_j}, \tag{5.61}$$

where q_j is the subgrid scale flux. A dynamic procedure can be used to determine σ_t, as the one described in the next section. In this work it is set to $\sigma_t = 1$, if not otherwise noted. In Section 6.3.1 the effect of the value of σ_t on the mean passive scalar is discussed in more detail.

The SGS turbulent viscosity ν_t is defined as

$$\nu_t = (C_S\Delta)^2|S| \tag{5.62}$$

where C_S is the Smagorinsky coefficient, Δ is the filter width and $|S| = \sqrt{2S_{ij}S_{ij}}$. For complex flows as the ones in this work, the most common value for the C_S constant is $C_S = 0.1$ [27].

Dynamic Smagorinsky model

Germano *et al.* [36] first proposed the procedure of dynamic SGS models. It can be implemented in any SGS model, and in this work it was used together with the Smagorinsky model.

Applying this procedure, the coefficient C_S of the Smagorinsky model is computed dynamically as the calculation progresses rather than having its values fixed a priori. The procedure is based on an algebraic identity between the SGS stresses at two different filtered levels and the resolved turbulent stresses, assuming that the behavior of the smallest resolved scales is very similar to the subgrid scales. The smallest resolved scales are sampled, and this information is used to model the subgrid scales. It is accomplished filtering again the filtered results $\overline{u_i}$ using a new test filter $\widehat{\cdot}$ with a wilder filter length ($\widehat{\Delta} > \Delta$), leading to a different SGS stress tensor

$$T_{ij} = \widehat{\overline{u_i u_j}} - \widehat{\overline{u}}_i \widehat{\overline{u}}_j. \tag{5.63}$$

The SGS stress tensors in Equations (5.55) and (5.63) are related by the Germano identity [36]

$$\begin{aligned}
\mathscr{L}_{ij} &= T_{ij} - \widehat{\tau_{ij}} && (5.64)\\
&= \left(\widehat{\overline{u_i u_j}} - \widehat{\overline{u}}_i \widehat{\overline{u}}_j\right) - \left(\widehat{\overline{u_i u_j} - \overline{u}_i \overline{u}_j}\right) && (5.65)\\
&= \widehat{\overline{u}_i \overline{u}_j} - \widehat{\overline{u}}_i \widehat{\overline{u}}_j. && (5.66)
\end{aligned}$$

Applying the definitions of the Smagorinsky model in the Equations (5.59) and (5.62) in (5.64) results in

$$\mathscr{L}_{ij} = -2C_S^2 \left(\widehat{\Delta}^2 |\widehat{S}| \widehat{S}_{ij} - \Delta^2 \widehat{|\overline{S}| \overline{S}_{ij}}\right) = -2C_S^2 M_{ij}. \tag{5.67}$$

The terms $\widehat{\overline{u_i u_j}} - \widehat{\overline{u}}_i \widehat{\overline{u}}_j$ and $M_{ij} = \overline{\Delta}^2 |\overline{S}| \overline{S}_{ij} - \widehat{\Delta}^2 \widehat{|\overline{S}| \overline{S}_{ij}}$ can be evaluated from the solution of the flow field, and the model coefficient C_S can be computed as

$$C_S^2 = \frac{1}{2} \frac{\mathscr{L}_{ij} \widehat{S}_{ij}}{M_{ij} \widehat{S}_{ij}}. \tag{5.68}$$

This definition, however, leads to numerical instabilities. Lilly [64] proposed a least squares procedure to evaluate C_S, which is more stable and is used in this work

$$C_S^2 = \frac{1}{2}\frac{\mathscr{L}_{ij}M_{ij}}{M_{ij}M_{ij}}. \tag{5.69}$$

The coefficient C_S evaluated using this equation is then applied in Equation (5.62) to calculate the SGS turbulent viscosity; this model is called dynamic Smagorinsky in the following chapters.

In the simulations presented in this work, the dynamic Smagorinsky model proved to be unstable, leading often to divergence of the whole simulation. Comparisons between the results using the standard and the dynamic versions of the Smagorinsky model can be seen in Section 6.3.1 for the free jet and in Section 7.2 for the jet in crossflow configuration.

5.4.3 Wall treatment

Applying LES to high Reynolds number flows makes the treatment of the wall region of critical importance. The requirement of placing the first grid point well within the viscous sub-layer ($y^+ < 1$) leads to computational requirements similar to DNS calculations in the same region, which increases the computational costs enormously. As an alternative, a model can be used to supply approximate statistics to the outer flow at a position away from the wall. One of the most widely known universal velocity profiles is Spalding's law of the wall [108], which satisfies the following conditions:

- passes through the point $y^+ = 0$, $u^+ = 0$;
- is tangential at this point to Equation (5.33) $u^+ = y^+$;
- is asymptotic at large y^+ to the logarithmic law, Equation (5.31) $u^+ = (1/\kappa)\ln(Ey^+)$;
- fits the experimental points at intermediate y^+ values,

where $\kappa = 0.4187$ and E = 9 are constants, $y^+ = yu_\tau/\nu$ is the dimensionless wall distance and $u^+ = \overline{u}/u_\tau$ is the dimensionless velocity. The equation is essentially a fit of the laminar, buffer and logarithmic regions of an equilibrium boundary layer

$$y^+ = u^+ + \frac{1}{E}\left[e^{\kappa u^+} - 1 - \kappa u^+ - \frac{1}{2}(\kappa u^+)^2 - \frac{1}{6}(\kappa u^+)^3\right]. \tag{5.70}$$

By substituting the known values of y and $\overline{u}$ next to the wall, a nonlinear equation for u_τ can be derived, which provide the wall shear. The resulting equation can be easily solved using an iterative procedure. Using such a universal velocity profile has the significant advantage of allowing the first grid point to be placed in the buffer or viscous regions ($y^+ < 30$) without the loss of accuracy usually associated with the limited validity of logarithmic profiles as the ones used in the RANS simulations. Provided the rest of the boundary layer and wall parallel directions are also adequately resolved, the simulation should recover a DNS boundary layer solution as the near-wall cell size tends toward zero, as noted by de Villiers [20].

5.4.4 LES of compressible flows

In compressible flows with variable density ρ, a mass-weighted Favre filtering is introduced according to

$$\overline{\rho}\widetilde{u_i}(x_i,t) = \int G(x_i - x_i')\rho u_i(x_i',t)dx_i', \tag{5.71}$$

where G is the filter function. The Favre filtered variables retain the properties of the filtered variables in incompressible flows, *i.e.*, the filtered residual is not zero in general

$$\widetilde{u_i'}(x_i,t) \neq 0. \tag{5.72}$$

Using the Favre filtering, the governing equations employed for compressible LES are derived [91]:

Overall continuity

$$\frac{\partial \overline{\rho}}{\partial t} + \frac{\partial \overline{\rho}\widetilde{u_i}}{\partial x_i} = 0, \tag{5.73}$$

Momentum

$$\frac{\partial \overline{\rho}\tilde{u}_j}{\partial t} + \frac{\partial \overline{\rho}\tilde{u}_i\tilde{u}_j}{\partial x_i} = -\frac{\partial \overline{p}}{\partial x_j} + \frac{\partial}{\partial x_i}\left(t_{ij} - \partial\overline{\rho}(\widetilde{u_i u_j} - \tilde{u}_i\tilde{u}_j)\right), \tag{5.74}$$

Chemical species

$$\frac{\partial \overline{\rho}\widetilde{Y}_k}{\partial t} + \frac{\partial \overline{\rho}\tilde{u}_i\widetilde{Y}_k}{\partial x_i} = \frac{\partial}{\partial x_i}\left(\overline{\rho}\overline{D}_k\frac{\partial \widetilde{Y}_k}{\partial x_i}\right) - \frac{\partial\overline{\rho}(\widetilde{u_i Y_k} - \tilde{u}_i\tilde{Y}_k)}{\partial x_i} + \overline{\dot{\omega}}_k, \quad \text{for } k = 1, N, \tag{5.75}$$

Enthalpy

$$\frac{\partial \overline{\rho}\widetilde{h}}{\partial t} + \frac{\partial \overline{\rho}\tilde{u}_i\widetilde{h}}{\partial x_i} = \frac{\partial}{\partial x_i}\left(\overline{\lambda}\frac{\partial \widetilde{T}}{\partial x_j}\right) + \overline{\frac{Dp}{Dt}} + \overline{\frac{\partial u_i}{\partial x_i}\left(\mu\frac{\partial u_i}{\partial x_j}\right)} - \frac{\partial\overline{\rho}(\widetilde{u_i h} - \tilde{u}_i\tilde{h}_k)}{\partial x_i}. \tag{5.76}$$

where t_{ij} is the constitutive relation between stress and strain rate for compressible flows, defined in Equation (5.39). The unclosed terms were modeled using the Smagorinsky SGS model [104], adapted to compressible flows. The unsolved SGS stresses $\overline{\rho}(\widetilde{u_i u_j} - \tilde{u}_i\tilde{u}_j) = \overline{\rho}\tau_{ij}$ for variable density flows must be adapted using Favre filtered quantities, incorporating the trace of the strain rate tensor to the SGS stresses:

$$\overline{\rho}\left(\tau_{ij} - \frac{1}{3}\tau_{kk}\delta_{ij}\right) = -2\mu_t\left(\tilde{S}_{ij} - \frac{1}{3}\tilde{S}_{kk}\delta_{ij}\right) \tag{5.77}$$

where μ_t is the subgrid scale turbulent viscosity and $\tilde{S}_{ij}$ is the strain rate tensor for the resolved scale. As in the incompressible formulation, only the anisotropic part of the subgrid scale stresses τ_{ij} are resolved by the subgrid scale model; the isotropic part is modeled together with the static pressure term. The SGS turbulent viscosity μ_t is defined as

$$\mu_t = \overline{\rho}(C_S\Delta)^2|\tilde{S}| \tag{5.78}$$

where Δ is the filter width and $|\tilde{S}| = \sqrt{2\tilde{S}_{ij}\tilde{S}_{ij}}$. The Smagorinsky coefficient C_S assumes the same value of 0.1 used in the incompressible formulation.

The subgrid scale turbulent fluxes are modeled using a subgrid scale turbulent Schmidt/Prandtl number σ_t

$$\overline{\rho}(\widetilde{u_i Y_k} - \tilde{u}_i \tilde{Y}_k) = -\frac{\mu_t}{\sigma_t}\frac{\partial \tilde{Y}_k}{\partial x_j}, \tag{5.79}$$

$$\overline{\rho}(\widetilde{u_i h} - \tilde{u}_i \tilde{h}) = -\frac{\mu_t}{\sigma_t}\frac{\partial \tilde{h}}{\partial x_j}, \tag{5.80}$$

where σ_t is equal to one, if not otherwise noted.

5.5 Turbulent combustion modeling

Combustion, as other chemical reactions, depend on the mixing at the molecular level of the reactants. After this condition is fulfilled, the chemical reaction can be initiated. In laminar flows, the molecular species diffusion is responsible for this mixing. If the flow is turbulent, on the other hand, fluid elements are convected by the turbulent fluctuations, which increases the mixing in comparison with a laminar flow.

One of the main challenges in modeling complex combustion systems is the description of the interaction between turbulent mixing and chemical reaction. In the literature, some different modeling approaches are known. The most frequently used reaction schemes can be divided into the following categories:

- **PDF models**
  These models are based on the probability density function (PDF) of the main variables of the chemical reaction. It is assumed that the time and length scales of mixing and reaction process are fully reflected in the shape of the PDF, *i.e.*, no restrictive assumptions regarding the chemical kinetics and the turbulent transport are made *a priori*.

- **Scalar dissipation rate**
  The scalar dissipation rate describes the relaxation of fluctuations due to molecular mixing. Models of this category are based on the assumption that the time scale of the reaction is always much smaller than the time scale of the turbulent mixing. It is assumed that the reaction rate depends only on the time scale controlled by the molecular mixture, which in turn is controlled by the turbulent length scale. Since the reaction rate is

proportional to the velocity in which the reactants come in contact in the flame front, the scalar dissipation rate appears directly or indirectly in the reaction source term.

- **Flame geometry**
  This modeling approach describes the kinematics of the flame front. It is assumed that the actual reaction front is thin compared with the integral length scale. To describe the topology of the reaction progress variable, there are different methods: equations for isosurfaces of the reaction progress (e.g. G-equation), balance equations for the flame front density and balance equations for the flame front curvature.

- **Turbulent burning velocity**
  Models based on correlations for the turbulent burning velocity describe the turbulence-reaction interaction based asymptotic time and length scales. Often the same restrictions used in the laminar case are used, with only one time scale for the heat release and the diffusive transport.

The sophisticated presumed Joint Probability Density Function (JPDF) has a long tradition in the Engler-Bunte-Institute, Division of Combustion Technology (EBI-VBT) and was the model of choice for the simulations in this work. The next session will present the details of this model.

5.6 Presumed JPDF

The presumed shape Joint Probability Density Function model (presumed JPDF or simply JPDF) [5, 69] has been developed in the Engler-Bunte-Institute, Division of Combustion Technology (EBI-VBT) by many colleagues [41, 38, 12, 120, 52, 33, 96, 34]. The implementation of the JPDF model in the OpenFOAM framework is due to Matthias Kern.

The basic idea of the JPDF model is that the interaction between turbulence and chemical reactions can be described by the probability density function of characteristic variables of the combustion system. The state of the mixing is determined using the mixing fraction f and the progress of the reaction using a reaction progress variable c. With this formalism, it is possible to simulate non-premixed, partially premixed and premixed flames using the same model.

The mixture fraction f is a passive scalar, and is defined as follows assuming hydrocarbon as fuel

$$f = \frac{(Z_C + Z_H) - (Z_C + Z_H)_{Ox}}{(Z_C + Z_H)_F - (Z_C + Z_H)_{Ox}} \tag{5.81}$$

where Z_C and Z_H represent the local mass fraction of carbon and hydrogen atoms, respectively, and the indices Ox and F represent pure oxidizer and pure fuel streams. The mixture fraction of the pure oxidant has a value of zero, while it has a value of unity in the pure fuel stream.

In this work, the reaction progress variable is defined using O_2 as characteristic variable:

$$c = \frac{Y_{O_2,local} - Y_{O_2,unburned}}{Y_{O_2,burned} - Y_{O_2,unburned}}. \tag{5.82}$$

where $Y_{O_2,local}$, $Y_{O_2,unburned}$ and $Y_{O_2,burned}$ represent the local mass fraction of O_2, the mass fraction in the unburned and in the completely burned states, respectively. The reaction progress variable is defined to assume a value of zero in the unburned mixture and unity in the completely burned mixture. It is a non-passive scalar, which is altered by the combustion progress.

The chemical reactions describing the combustion can be solved using several idealized systems, including flamelets, premixed laminar flames, non-premixed laminar flames and plug flow reactors. In this work, the plug flow reactor (PFR) was employed. The popular flamelet generated manifolds (FGM) model [117, 119] is very similar to the JPDF model using flamelets or premixed laminar flames to solve the chemical reactions.

A detailed chemical mechanism was used to model the combustion. The fuel used in this work is methane, and the chemical reactions have been described with the 53 species and 325 reactions of the GRI 3.0 mechanism [105].

For the range of mixture fractions being studied, mixtures of oxidizer and fuels streams are defined as educt of plug flow reactors and computed using the Cantera software [37]. For each mixture fraction f, the mass fraction of the different species change along the reactor due to the reaction between fuel and oxidizer. The independent variable of the PFR is the time, which can be transformed using the reaction progress variable c defined in Equation (5.82); this process is depicted in Figure 5.2. The line plots on the top show the time evolution of temperature and mass fractions of O_2 and CH_4. The reaction

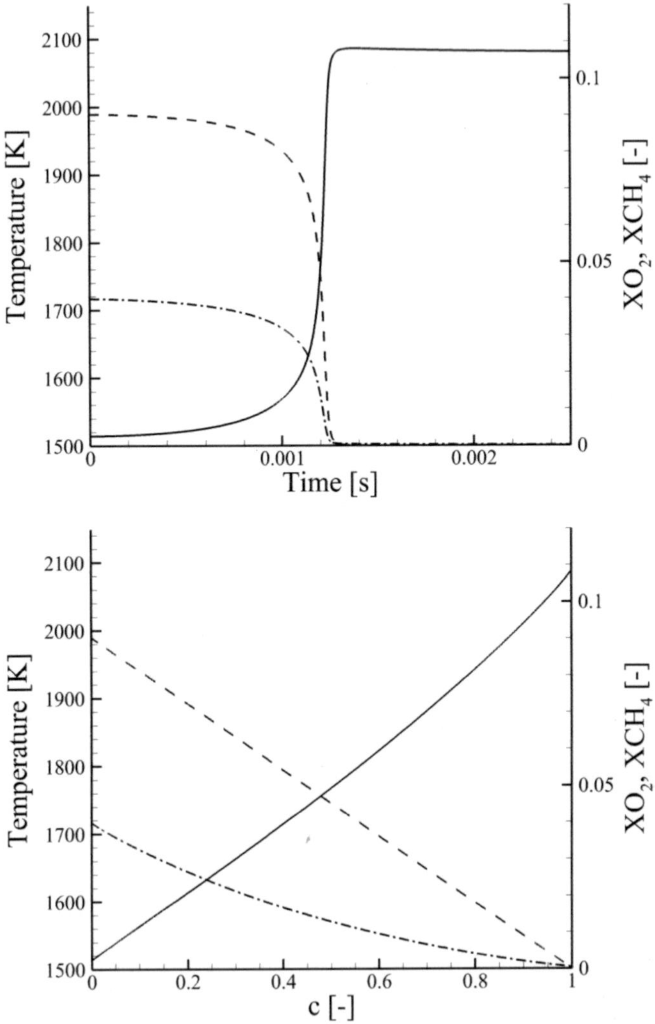

Figure 5.2: Line plots of the same results of a plug flow reactor using time (top) and reaction progress variable c (bottom) as independent variables. Temperature, solid line; mass fraction of O_2, dashed line; mass fraction of CH_4, dash-dot line.

zone is clearly determined by the steep temperature rise with simultaneous consumption of fuel (CH_4) and oxidizer (O_2). In this example, the oxidizer is vitiated air and the fuel is CH_4 diluted in air. The mixture is stoichiometric, so all O_2 is consumed by the reaction with CH_4. Applying the definition of the reaction progress variable c and using this variable as independent variable leads to the line plots shown in the bottom of Figure 5.2. The O_2 mass fraction now varies linearly, which is a consequence of using O_2 as the characteristic variable in the definition of c. The variations of the temperature and the mass fraction of CH_4 become also smoother.

The next step is to use the mixture fraction f and the reaction progress variable c as independent variables to tabulate all other variables of interest: mixture density, mass fractions of all species of interest, reaction rate of c ($\dot{\omega}_c$), mixture viscosity, mixture heat and mass diffusivity, among others. The result is a two-dimensional table, depicted in Figure 5.3, showing the reaction rate of the reaction progress variable $\dot{\omega}_c$ as a function of f and c. The interested reader can find the detailed boundary conditions in Table 8.2 with equivalence ratio 1.02.

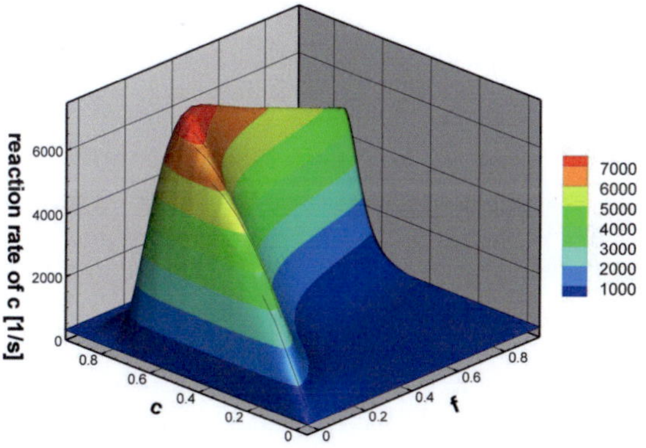

Figure 5.3: Example of distribution of reaction rate of c ($\dot{\omega}_c$) with mixture fraction f and reaction progress variable c as independent variables.

The results in the two-dimensional table are still laminar, without the influence of turbulence. To consider the turbulence, PDFs of each independent variable are computed. The β probability density function is widely used in turbulent combustion to define scalar distributions [91]; it depends on two moments only, the mean value and the variance. Using the mixture fraction f as an example, the β PDF assumes the form

$$P(f) = \frac{\Gamma(\alpha+\beta)}{\Gamma(\alpha)\Gamma(\beta)} f^{\alpha-1}(1-f)^{\beta-1}, \tag{5.83}$$

where P is the probability, Γ is the gamma function, and the parameters α and β are related through

$$\alpha = \widetilde{f}\gamma \tag{5.84}$$

$$\beta = (1-\widetilde{f})\gamma \tag{5.85}$$

where $\widetilde{f}$ is the mean value of f and γ is defined as

$$\gamma = \frac{\widetilde{f}(1-\widetilde{f})}{\widetilde{f''^2}} - 1 \tag{5.86}$$

where $\widetilde{f''^2}$ is the variance of f.

Statistical independence of the mixture fraction f and the reaction progress variable c is assumed, and the joint probability is calculated multiplying the marginal PDFs [93, 91]

$$P(f,c) = P(f)P(c). \tag{5.87}$$

Unfortunately, it is very difficult to confirm this assumption in practice. The statistical independence of f and c is weak, because the mixture fraction and the reaction progress variable are clearly linked in flames. Nevertheless, the joint PDF calculated this way has been extensively used in the literature [91, 119, 23, 76] and in previous works published by the Engler-Bunte-Institute, Division of Combustion Technology (EBI-VBT) [120, 52], with good results.

Applying this formalism, a lookup table with values including the influence of the turbulence can be calculated. The four independent values of this table are the mean values and variances of f and c. Employing the β PDF, the marginal PDFs of f and c are calculated, and using Equation (5.87) the joint probability is defined. The lookup table is then populated with the mean values of various variables, including density, species mass fraction and reaction rates, which are calculated using their laminar values and the joint probability.

The next step is to calculate the mean values and the variances of f and c in the CFD code, defining the progress of the mixing and of the reaction in every point of the simulation. The mean values of other variables can then be read from the lookup table, for example the mean density $\overline{\rho}$, mean reaction rate of c $\overline{\dot{\omega}}_c$, mean source term of d $\overline{\dot{\omega}_c' d'}$, and other variables of interest.

In this version, the JPDF model does not account for heat losses, which are present in the combustion system presented in Chapter 8. Heat losses can affect the flame geometry significantly, influencing the agreement between the measurements and the simulations. An extension of the JPDF that considers heat losses has been already developed [120] and should be used in future simulations.

The proper transport equations for the mixture fraction f, the reaction progress variable c and their variances depend on the type of simulation being conducted. For simulations using RANS (Reynolds-averaged Navier Stokes) turbulence models, the transport equations are described in the next section. Transport equations for large eddy simulations (LES), in which the equations are not averaged but filtered, are described in Section 5.6.2.

5.6.1 RANS equations

For simulations using RANS turbulence models, the transport equations for the mean mixture fraction $\widetilde{f}$ and its variance $\widetilde{g} = \widetilde{f''^2}$ are the following:

$$\frac{\partial \overline{\rho}\widetilde{f}}{\partial t} + \frac{\partial \overline{\rho}\widetilde{u}_i\widetilde{f}}{\partial x_i} - \frac{\partial}{\partial x_i}\left(\overline{\rho} D_{Eff}\frac{\partial \widetilde{f}}{\partial x_i}\right) = 0 \tag{5.88}$$

$$\begin{aligned}\frac{\partial \overline{\rho}\widetilde{g}}{\partial t} + \frac{\partial \overline{\rho}\widetilde{u}_i\widetilde{g}}{\partial x_i} - \frac{\partial}{\partial x_i}\left(\overline{\rho} D_{Eff}\frac{\partial \widetilde{g}}{\partial x_i}\right) = C_{g,1}\overline{\rho} D_{Eff}\left|\frac{\partial \widetilde{f}}{\partial x_i}\right|^2 \\ -C_{g,2}\overline{\rho}\frac{\varepsilon}{k}\widetilde{g}\end{aligned} \tag{5.89}$$

where D_{Eff} is the effective diffusion coefficient, including the effect of the turbulence, $C_{g,1} = 2.8$ and $C_{g,2} = 2$.

The mean reaction progress variable $\widetilde{c}$ is a non-passive scalar, and its transport equation includes a reaction source term. The same applies to its variance $\widetilde{d} = \widetilde{c''^2}$:

$$\frac{\partial \overline{\rho}\widetilde{c}}{\partial t} + \frac{\partial \overline{\rho}\widetilde{u}_i\widetilde{c}}{\partial x_i} - \frac{\partial}{\partial x_i}\left(D_{Eff}\frac{\partial \widetilde{c}}{\partial x_i}\right) = \overline{\dot{\omega}}_c \tag{5.90}$$

$$\begin{aligned}\frac{\partial \overline{\rho}\widetilde{d}}{\partial t} + \frac{\partial \overline{\rho}\widetilde{u}_i\widetilde{d}}{\partial x_i} - \frac{\partial}{\partial x_i}\left(\overline{\rho}D_{Eff}\frac{\partial \widetilde{d}}{\partial x_i}\right) &= C_{d,1}\overline{\rho}D_{Eff}\left|\frac{\partial \widetilde{c}}{\partial x_i}\right|^2 \\ &\quad -C_{d,2}\overline{\rho}\frac{\varepsilon}{k}\widetilde{d} + 2\cdot\overline{\dot{\omega}_c' d'}\end{aligned} \tag{5.91}$$

where $C_{d,1} = 2.8$ and $C_{d,2} = 2$. The source terms are defined as:

$$\overline{\dot{\omega}}_c = \overline{\rho}\int_1^0 \frac{\dot{\omega}}{\rho}P(f,c)dcdf \tag{5.92}$$

$$\overline{\dot{\omega}_c' d'} = \overline{\rho}\int_1^0 \frac{(\dot{\omega}-\overline{\dot{\omega}})(c-\tilde{c})}{\rho}P(f,c)dcdf \tag{5.93}$$

where $P(f,c)$ is the joint probability of f and c defined in Equation (5.87).

5.6.2 LES equations

The transport equations for the LES framework are derived using the mass-weighted Favre filtering operation, defined in Equation (5.71). The transport equation for the filtered mixture fraction $\widetilde{f}$ is:

$$\frac{\partial \overline{\rho}\widetilde{f}}{\partial t} + \frac{\partial \overline{\rho}\widetilde{u}_i\widetilde{f}}{\partial x_i} - \frac{\partial \overline{\rho}(\widetilde{u_i f} - \widetilde{u}_i\widetilde{f})}{\partial x_i} = 0 \tag{5.94}$$

As in the RANS equation, the transport equation for the filtered reaction progress variable $\widetilde{c}$ includes a reaction source term:

$$\frac{\partial \overline{\rho}\widetilde{c}}{\partial t} + \frac{\partial \overline{\rho}\widetilde{u}_i\widetilde{c}}{\partial x_i} - \frac{\partial \overline{\rho}(\widetilde{u_i c} - \widetilde{u}_i\widetilde{c})}{\partial x_i} = \overline{\dot{\omega}}_c \tag{5.95}$$

where the source term is defined as:

$$\overline{\dot{\omega}}_c = \overline{\rho} \int_1^0 \frac{\dot{\omega}}{\rho} P(f,c) dc df \tag{5.96}$$

where $P(f,c)$ is the joint probability of f and c as defined in Equation (5.87). The unsolved subgrid scale stresses $\overline{\rho}(\widetilde{u_i f} - \widetilde{u}_i \widetilde{f})$ and $\overline{\rho}(\widetilde{u_i c} - \widetilde{u}_i \widetilde{c})$ are modeled by subgrid scale models already described in Section 5.4.

In LES the variances of $\widetilde{f}$ and $\widetilde{c}$ can be calculated using algebraic relations, *i.e.*, without solving transport equations. The algebraic relations are also described in Section 5.4.

Subgrid scale variance

The presumed JPDF model requires the evaluation of the variances of selected variables. In RANS context, transport equations are solved for these variances. In LES, the variances can be calculated using an algebraic relation, an approach comparable to that employed by Smagorinsky to approximate the subgrid scale turbulent kinetic energy. The SGS variance of the mixture fraction f and the reaction progress variable c are given by

$$\widetilde{f''^2} = C_f \Delta^2 \left(\frac{\partial \tilde{f}}{\partial x_i} \right)^2 \tag{5.97}$$

$$\widetilde{c''^2} = C_c \Delta^2 \left(\frac{\partial \tilde{c}}{\partial x_i} \right)^2 \tag{5.98}$$

where the constants C_f and C_c assume the value 0.01. Similar procedures have been employed by Branley and Jones [8], Vreman *et al.* [119] and Olbricht *et al.* [82], with good results.

5.7 Flow Solver

The conservation equation, turbulence models and wall treatments presented in the preceding sections constitute the basis of CFD; however, they are a comparatively small part of a functional CFD solution method. The methodology includes the discretization of the governing equations, the pressure-velocity coupling, and the numerical solution of the resultant matrices along with many additional functionalities. Fortunately, most of these methodologies have been extensively covered in various other publications, so that only the portions that relate directly to this work are reproduced here for completeness.

The simulations were performed using the open-source CFD toolbox OpenFOAM version 2.0 [85]. The monographs of Jasak [46], one of the early developers, and de Villiers [20], focused on LES methods, contain much information about the implementation of the different methodologies in OpenFOAM and serve as the foundation for the next sections.

5.7.1 Discretization of the computational domain

The purpose of the discretization is to transform the differential equations into a corresponding system of algebraic equations, which can be numerically solved. The discretization of the computational domain can be subdivided into spatial and temporal discretization. Spatial discretization defines the solution domain as a collection of well-defined control volumes that fill and bound the space (or region of space) of interest. The centroid of each control volume defines a computational point. OpenFOAM uses the Finite Volume method (FVM) of spatial discretization. It is based on the integral form of the governing equations discretized over each control volume. The basic quantities, such as mass and momentum, are therefore conserved at the discrete level. The software accepts unstructured grids with control volumes of any polyhedral shape, with any number of neighbors. The variable arrangement is collocated, in which all dependent variables share the same control volumes, in contrast to a staggered arrangement, in which the scalar variables (pressure, density, etc.) are stored in the cell centers, whereas the velocity is located at the cell faces. The equations are solved one at a time, in a segregated way, with explicit inter-equation coupling.

For transient simulations, the time interval is divided into a finite number of time steps. The solution is obtained by marching forward from an initial condition, with a specified time step size Δt.

A typical control volume is shown in Figure 5.4. The computational point P is located at the centroid of the control volume, such that

$$\int_{V_P} (x_i - x_{i,P}) dV = 0. \tag{5.99}$$

The control volume is bounded by a set of flat faces and each face is shared with only one neighboring control volume. OpenFOAM accepts control volumes of any topology, including tetrahedra, prisms and hexahedra.

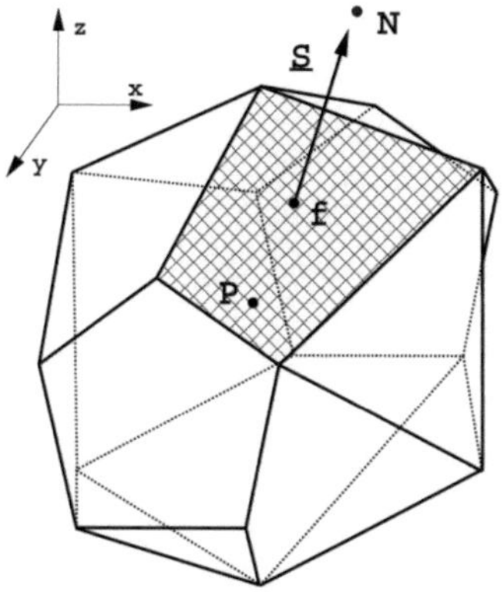

Figure 5.4: Control volume. Reproduced from [46].

Convection term

The discretization of the convection term is the most difficult to solve. It determines the value of a variable ϕ on the face of the finite volume elements from the values in the cell centers.

Assuming that ϕ varies linearly between P and N, as shown in Figure 5.5, the face centered value can be found from a simple interpolation between the cell values

$$\phi_f = f_x \phi_P + (1 - f_x)\phi_N, \tag{5.100}$$

where the interpolation factor f_x is defined as the ratio of distances $\overline{fN}$ and $\overline{PN}$:

$$f_x = \frac{\overline{fN}}{\overline{PN}}. \tag{5.101}$$

This practice is commonly known as *Central Differencing* (CD) and is second order accurate on unstructured meshes. This scheme has some disadvantages, as its tendency to produce unphysical oscillations in the solution when the convection term strongly dominates the rest of the system [87], thus violating the boundedness of the solution.

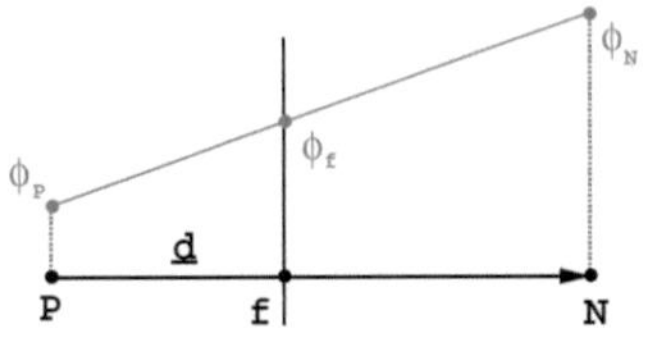

Figure 5.5: Face interpolation. Reproduced from [46].

An alternative discretization scheme that guarantees boundedness is the *Upwind Differencing* (UD), in which the face value is determined according to the flow direction

$$\phi_f = \begin{cases} \phi_f = \phi_P & \text{for } F \geq 0 \\ \phi_f = \phi_N & \text{for } F < 0, \end{cases} \tag{5.102}$$

where $F = \rho u$ represent the convective mass flux per unit area. The boundedness of the solution is guaranteed at the expense of the accuracy, by implicitly introducing numerical diffusion into the system. It is only first order accurate, thus violating the order of accuracy of the discretization.

In an attempt to preserve both boundedness and accuracy of the solution, *Blended Differencing* is introduced. It is a linear combination of UD, Equation (5.102) and CD, Equation (5.100)

$$\phi_f = (1-\gamma)(f_x)_{UD} + \gamma(f_x)_{CD}, \tag{5.103}$$

where the blending factor γ, $0 \leq \gamma \leq 1$, determines how much numerical diffusion will be introduced. There are many attempts in finding an acceptable compromise between accuracy and boundedness, as "‘streamwise-upwind schemes"’ [97, 98], higher-order upwind schemes (*e.g.* QUICK [62]) and flux-limited schemes (*e.g.* [6]). Flux limiting creates differencing schemes that are higher than first-order accurate, but without the spurious oscillations associated with the classical second-order schemes.

The *Gamma* scheme developed by Jasak [46] uses Central Differencing (CD) in the bulk of the domain. In the regions where the boundedness of ϕ is violated, the model prescribes Upwind Differencing (UD) in order to guarantee boundedness. To avoid perturbations from the switching between the schemes, a smooth blending between UD and CD is introduced.

The blending between UD and CD is based on the normalized variable $\widetilde{\phi_C}$, using the nodal values of ϕ at points U, C and D (see Figure 5.6). It is defined as

$$\widetilde{\phi_C} = \frac{\phi_C - \phi_U}{\phi_D - \phi_U}. \tag{5.104}$$

The blending is smooth over the interval $0 < \widetilde{\phi_C} < \beta_m$, where β_m is a constant

$$\gamma = \frac{\widetilde{\phi_C}}{\beta_m}. \tag{5.105}$$

The limit behavior of the function is such that

$$\begin{aligned} \widetilde{\phi_C} &= 0 && \Rightarrow \gamma = 0 \text{ (Upwind Differencing)} \\ \widetilde{\phi_C} &= \beta_m && \Rightarrow \gamma = 1 \text{ (Central Differencing).} \end{aligned} \tag{5.106}$$

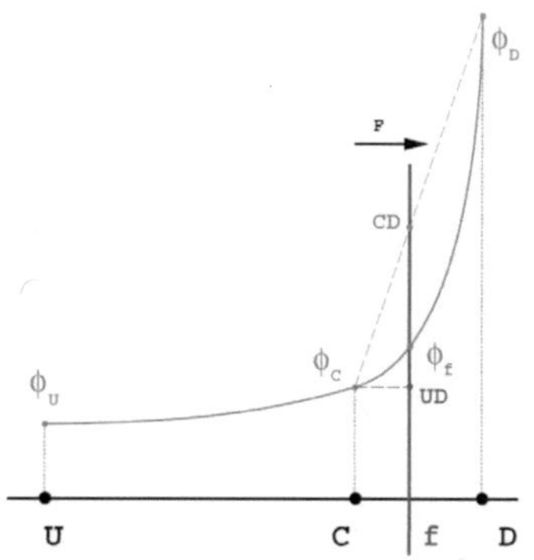

Figure 5.6: Variation of ϕ around the face f. The values of ϕ using central differencing (CD) and upwind differencing (UD) are indicated. Reproduced from [46].

The larger the value of β_m, the more blending will be introduced. For better resolution of the numerical scheme, the value of β_m should be kept as low as possible. The useful range of β_m is

$$0.1 < \beta_m < 0.5. \tag{5.107}$$

For the steady state RANS simulations, the value of β_m was kept to 0.5 to improve the convergence of the calculations, as recommended by Jasak [46].

The numerical diffusion is particularly problematic in LES context, since the contribution of the modeled turbulent diffusivity is typically very small, so that even modest numerical diffusion can induce large inaccuracies. For that purpose, the Filtered Linear scheme was developed by Henry Weller for OpenFOAM [85]. It is a low dissipation second order scheme based on Central Differencing, in which the filtering removes the staggering caused by pressure-velocity decoupling by introducing small amounts of Upwind Differencing. It addresses the problem of staggering without adversely affecting the LES statistics.

Although appropriate for the discretization of the velocity in LES context, the Filtered Linear scheme remains unbounded, which can have severe effects on strictly bounded variables as the passive scalar. For that reason, the Gamma scheme, being bounded and stable, was employed for the passive scalar in this work. The blending factor was kept to the minimum value of 0.1 proposed by Jasak [46], limiting the adverse effects of the added numerical diffusion.

Diffusion and source terms

The discretization of the diffusion and source terms is less complex than the discretization of the convection term described above. The only problem faced while discretizing the diffusion term is the non-orthogonality of the grid, whose effects have to be modeled. Following the assumption of linear variation of ϕ and using the divergence theorem, the integration of the diffusion term (*e.g.* from Equation (2.3)) can be represented as the sum of fluxes across the faces of the control volume

$$\int_{V_P} \frac{\partial}{\partial x_j}\left(D\frac{\partial \phi}{\partial x_j}\right) dV = \sum_f \vec{S}\cdot\left(D\frac{\partial \phi}{\partial x_j}\right)_f . \tag{5.108}$$

where $\vec{S}$ is the surface vector of the face f. Figure 5.7 shows schematically a computational cell, with the face f between the points P and N. If the mesh is orthogonal, the distance vectors $\vec{d}$ and $\vec{S}$ are parallel, and the following expression is valid

$$\vec{S}\cdot\left(\frac{\partial\phi}{\partial x_j}\right)_f = |\vec{S}|\frac{\phi_N-\phi_P}{|\vec{d}|}. \tag{5.109}$$

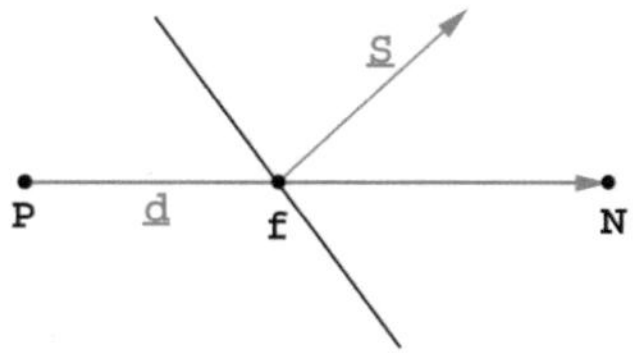

Figure 5.7: Vectors $\vec{d}$ and $\vec{S}$ in a non-orthogonal mesh. Reproduced from [46].

Unfortunately, in practice orthogonal meshes are more an exception than a rule. To account for the non-orthogonality, the product of the surface vector with the gradient of ϕ is split into two parts

$$\vec{S}\cdot\left(\frac{\partial\phi}{\partial x_j}\right)_f = \vec{\Delta}\cdot\left(\frac{\partial\phi}{\partial x_j}\right)_f + \vec{k}\cdot\left(\frac{\partial\phi}{\partial x_j}\right)_f. \tag{5.110}$$

where the first term in the right-hand side represents the contribution considering orthogonal mesh and the second term the non-orthogonal correction. The two vectors introduced in Equation (5.110), $\vec{\Delta}$ and $\vec{k}$, are related to $\vec{S}$ in the following matter

$$\vec{S} = \vec{\Delta}+\vec{k}. \tag{5.111}$$

Many different approaches can be employed for the decomposition. In the *minimum correction approach* (see Figure 5.8), the non-orthogonal correction is kept to a minimum, by making $\vec{\Delta}$ and $\vec{k}$ orthogonal

$$\vec{\Delta} = \frac{\vec{d}\cdot\vec{S}}{\vec{d}\cdot\vec{d}}\cdot\vec{d}, \tag{5.112}$$

with $\vec{k}$ calculated using Equation (5.111). As the non-orthogonality increases, the contribution from ϕ_P and ϕ_N decreases.

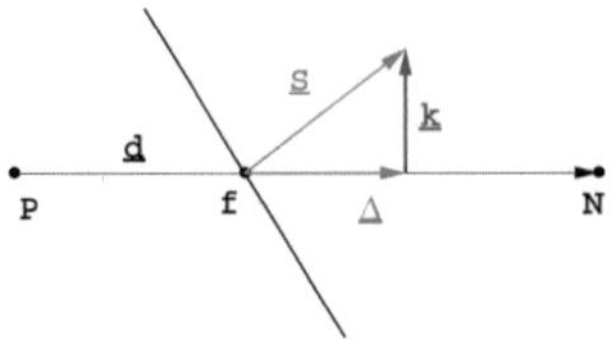

Figure 5.8: Non-orthogonality treatment in the *minimum correction* approach. Reproduced from [46].

To keep the contribution of ϕ_P and ϕ_N the same irrespective of the non-orthogonality, the *orthogonal correction approach* can be employed (see Figure 5.9), where $\vec{\Delta}$ is defined as

$$\vec{\Delta} = \frac{\vec{d}}{|\vec{d}|} \cdot |\vec{S}|. \tag{5.113}$$

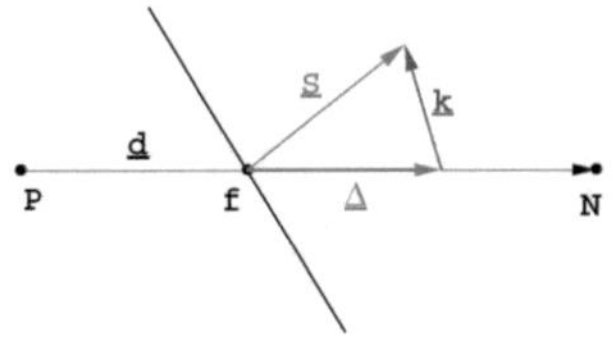

Figure 5.9: Non-orthogonality treatment in the *orthogonal correction* approach. Reproduced from [46].

In the *over-relaxed approach* (see Figure 5.10), the contribution of ϕ_P and ϕ_N increases with an increase in non-orthogonality

$$\vec{\Delta} = \frac{\vec{d}}{\vec{d} \cdot \vec{S}} \cdot |\vec{S}|^2. \tag{5.114}$$

The final form of the discretized diffusion term is the same for all three approaches. Taking in advantage that $\vec{d}$ and $\vec{\Delta}$ are parallel, the orthogonal part of Equation (5.110) is discretized as

$$\vec{\Delta} \cdot \left(\frac{\partial \phi}{\partial x_j} \right)_f = |\vec{\Delta}| \frac{\phi_N - \phi_P}{|\vec{d}|}. \tag{5.115}$$

and Equation (5.110) can be written as

$$\vec{S} \cdot \left(\frac{\partial \phi}{\partial x_j} \right)_f = |\vec{\Delta}| \frac{\phi_N - \phi_P}{|\vec{d}|} + \vec{k} \cdot \left(\frac{\partial \phi}{\partial x_j} \right)_f. \tag{5.116}$$

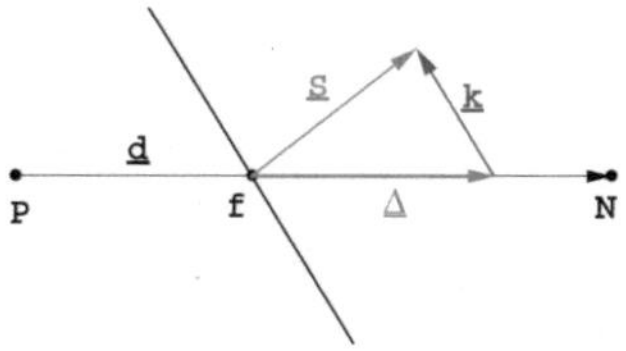

Figure 5.10: Non-orthogonality treatment in the *over-relaxed* approach. Reproduced from [46].

Jasak [46] draws the conclusion that the *over-relaxed* approach is the most robust, convergent and computationally efficient. This approach was used in all simulations in this work.

All terms of a transport equation that cannot be written as convection, diffusion or temporal contributions are loosely classified as source terms [46]. Before the discretization, the functions acting as source terms (S_ϕ) are linearized

$$S_\phi(\phi) = S_u + S_p \phi. \tag{5.117}$$

where S_u and S_p can also depend on ϕ. The source terms are subsequently integrated

$$\int_{V_P} S_\phi(\phi) dV = S_u V_P + S_p V_P \phi_P. \tag{5.118}$$

More details about the treatment of source terms can be found in Jasak [46] and Patankar [87].

Temporal discretization

Steady-state problems have a solution that is not a function of time. Steady-state RANS simulations are of this type. If solving a single steady-state equation, the solution can be obtained in a single step. However, as fluid flow problems are a non-linear system of coupled equations, it is necessary to solve the system in an iterative manner. In order to speed up the convergence, an implicit formulation is preferred. The convergence of the iterative procedure can be improved through under-relaxation, which is applied to each equation individually.

The different temporal discretization schemes implemented in OpenFOAM are described extensively in the monographs of Jasak [46] and de Villiers [20]. The most common methods of temporal discretization are summarized below.

For unsteady simulations as LES and URANS, the temporal discretization involves the integration of every term in the differential equations over a time step Δt.

The most simple temporal discretization method implemented in OpenFOAM is the *Euler Implicit*. This method expresses the face values in terms of the new time-level cell values

$$\phi_f = f_x \phi_P^n + (1 - f_x)\phi_N^n, \tag{5.119}$$

and is unconditionally stable. It guarantees boundedness of the solution; however, it is only first order accurate.

The *Crank-Nicholson* method is second order accurate, but does not guarantee boundedness of the solution. The time derivative is calculated as

$$\left(\frac{\partial \phi}{\partial t}\right) = \frac{\phi^n - \phi^{n-1}}{\Delta t}, \tag{5.120}$$

The Crank-Nicholson method is unconditionally stable; however, it requires inner-iterations during each time step. Coupled with the memory overhead due to the large number of stored variables, this means it is more expensive compared to the Backward Differencing scheme described below.

Backward Differencing is second-order accurate in time and still neglects the temporal variation of the face values. In order to achieve this, the Backward Differencing in time uses three time levels, $n-2=t-\Delta t$, $n-1=t$ and $n=t+\Delta t$. The temporal derivative at the time n is calculated as

$$\left(\frac{\partial \phi}{\partial t}\right)^n = \frac{\frac{3}{2}\phi^n - 2\phi^{n-1} + \frac{1}{2}\phi^{n-2}}{\Delta t}, \tag{5.121}$$

Again, the boundedness of the solution is not guaranteed. This is the method used in this work for the unsteady simulations.

5.7.2 Pressure-Velocity coupling

The SIMPLE pressure-velocity coupling procedure by Patankar and Spalding [88] was used for the steady-state simulations. For unsteady simulations, the PISO procedure proposed by Issa [45] was used for pressure-velocity coupling. Both schemes use an equation for pressure correction that is derived from the continuity equation and solved instead of it.

The SIMPLE (Semi-Implicit Method for Pressure-Linked Equations) algorithm is formulated to take advantage of the fact that it is not necessary to fully resolve the linear pressure-velocity coupling if the simulation is in steady state, as the changes between consecutive iterations are no longer small. Since the effective time step is much larger, the non-linearity of the system becomes more important. The algorithm follows these steps:

1. The momentum equations are solved, resulting in an approximation of the velocity field. The pressure gradient term is calculated using the pressure from the previous iteration. The equation is under-relaxed with the under-relaxation factor α_u.

2. The mass flux at the cell faces is computed.

3. The pressure correction equation is solved in order to obtain the new pressure distribution. The equation is under-relaxed with the under-relaxation factor α_p.

4. The mass flux at the cell faces is corrected.

5. On the basis of the new pressure field, the velocities are corrected.

In the SIMPLE algorithm, the pressure and the momentum are under-relaxed. The recommended values of under-relaxation factors are 0.2 for α_p and 0.8 for α_u [46].

The PISO (Pressure-Implicit with Splitting of Operators) pressure-velocity coupling scheme consists of an implicit momentum predictor followed by a series of pressure correction solutions and explicit velocity corrections. The loop is repeated until a pre-determined tolerance is reached. The main differences from the SIMPLE algorithm are that no under-relaxation is applied and that the momentum corrector step is performed more than once. The algorithm follows these steps:

1. The momentum equations are solved, resulting in an approximation of the velocity field. The pressure gradient term is calculated using the pressure from the previous iteration.

2. The mass flux at the cell faces is computed.

3. The pressure correction equation is solved in order to obtain the new pressure distribution. The equation is under-relaxed with the under-relaxation factor α_p.

4. The mass flux at the cell faces is corrected.

5. On the basis of the new pressure field, the velocities are corrected.

6. The algorithm is repeated from item 2 for the prescribed number of times.

7. The time step is increased and the algorithm is repeated from item 1.

The major restriction of the PISO scheme is that the Courant number of each cell has to be smaller than one

$$\text{Courant number} = \frac{u \cdot \Delta t}{\Delta x} \leq 1 \tag{5.122}$$

This is a really limiting constraint. For the flows of interest in this work, it limits the effective Δt to values of microseconds.

5.7.3 Boundary conditions

Boundary conditions are one of the most critical parameters of CFD simulations and should not be underestimated. The quality of the simulation results is directly coupled with the quality of the boundary conditions.

The physical boundary conditions of the problem have to be translated into a numerical boundary condition for the simulation. There are two basic types of numerical boundary conditions. The Dirichlet (or fixed value) boundary condition prescribes the value of the variable on the boundary. The von Neumann boundary condition, on the other hand, prescribes the gradient of the variable normal to the boundary.

Examples of physical boundary conditions are walls, inlet and outlet conditions for fluid flow problems, and adiabatic or fixed temperature boundaries for heat transfer problems, among others. Each of these conditions is associated with a set of numerical boundary conditions on each of the variables that are being calculated.

The prescription of turbulence properties at the boundaries can be especially complex. Within the classical RANS-approach, mean velocities and the transported turbulence quantities such as turbulent kinetic energy k and its rate of dissipation ε have to be set. These quantities can be determined with moderate experimental effort or may be estimated.

In LES context, on the other hand, the turbulent fluctuation is no longer described by simple, steady, transported quantities. Instead, the turbulent fluctuations are an integral part of the transient, fluctuating velocity field. It creates a problem at the inflow, where these fluctuations must be prescribed. This leads to a vicious circle, since the transient flow-field on the inflow must be known prior to the simulation. To solve this problem, a boundary condition based on the work of Klein *et al.* [53] was implemented in OpenFOAM and is described below.

Simulations of compressible flows using LES suffer from pressure wave reflections from the computational domain boundaries. Steady state simulations are not interested in the unsteady behavior of the flow or of the boundaries as long as the final steady state is reached. In these simulations, acoustic waves are eliminated by the numerical dissipation of the interpolation schemes employed. LES uses interpolation schemes with less numerical dissipation in order to describe the turbulence and all unsteady flow structures with higher accuracy, with the side effect of not damping acoustic waves. Non-reflecting boundary condition is used to control the acoustic waves in the computational domain and is described below.

Turbulent inlet boundary condition for LES

The simplest way to generate turbulent inflow data for a LES is to take a mean velocity profile and superimposed random fluctuations. Klein *et al.* [53] shows that this is not a very good method. Due to a lack of energy in the low wave number range, the fluctuations are almost immediately damped to zero, and the flow becomes laminar right after the inflow.

The algorithm proposed by Klein *et al.* [53] generates synthetic turbulent fluctuations from random noise. The first step is to create a velocity field with homogeneous, isotropic turbulence with prescribed integral length scale and energy spectra, which is described below. If cross-correlations between the different velocity components are desired, the method proposed by Lund *et al.* [70] can be used.

An equidistant structured grid of mesh spacing Δ is defined with logical coordinates i, j, k in the x, y, z directions, respectively. The coordinate x is aligned with the direction of the bulk velocity U_b, and can be replaced by the time using Taylor's hypothesis ($t = x/U_b$).

The algorithm creates and stores three independent fields of random numbers r^α, representing each component of the flow velocity ($\alpha = [u, v, w]$). The cross-section of the created fields is larger than the inflow area, in order to accommodate the filter width at the borders. The streamwise extension of the fields should be larger than $U_b \Delta t$, where Δt is the time step of the simulation. The low-pass filter $B_{i,j,k}$ is applied to the random fields r^α for every grid point $[I, J, K]$, resulting in the filtered fields $R^\alpha_{I,J,K}$

$$R^\alpha_{I,J,K} = \sum_{i=-N}^{N} \sum_{j=-N}^{N} \sum_{k=-N}^{N} B_{i,j,k} r^\alpha_{I+i,J+j,K+k}. \tag{5.123}$$

It is the low-pass filter $B_{i,j,k}$ that is responsible for the recovery of the integral length scale L by filtering the random noise. In the case of fully developed, homogeneous turbulence, the autocorrelation function R_{uu} assumes the form

$$R_{uu} = \exp\left(\frac{-\pi r^2}{4L^2}\right) \tag{5.124}$$

where r is the radius. Using a normalized length scale $n = L/\Delta$, the function can be written in discretized form

$$R_{uu}(i,j,k) = \exp\left(\frac{-\pi(i^2+j^2+k^2)}{4n^2}\right) \text{ with } \Delta^2(i^2+j^2+k^2) = r^2 \tag{5.125}$$

By the convolution of three one-dimensional filters, the three-dimensional filter $B_{i,j,k}$ can be obtained

$$B_{i,j,k} = b_i b_j b_k \tag{5.126}$$

where the one-dimensional filters are determined as

$$b_k \approx \widetilde{b}_k / \left(\sum_{j=-N}^{N} \widetilde{b}_j^2\right)^{\frac{1}{2}} ;\ \widetilde{b}_k = \exp\left(\frac{-\pi s^2}{2n^2}\right) \tag{5.127}$$

The filtered fields are then scaled with the mean velocity and the fluctuation scale to generate the velocity fields u^α.

One drawback of the algorithm as proposed by Klein *et al.* [53] is the spurious pressure fluctuations generated at the inlet. The main source of the fluctuations is that the turbulent velocity field generated by the Klein's method leads to small fluctuations of the mass flow. The solution is to apply a correction to the velocity field in every time step, to ensure that the mass flow remains constant. With this procedure, the amplitude of the pressure fluctuation is significantly attenuated. The remaining pressure fluctuations are related to the fact that the turbulent velocity field generated by the Klein's method is not divergence free, *i.e.*, it does not satisfy continuity. This fraction of the pressure fluctuations has its origin in the effort of the flow solver in forcing the velocity field to follow the continuity. Nevertheless, the amplitude of these pressure fluctuations is small and they are dissipated in a short distance, having a limited effect in the simulation as a whole.

Non-reflecting outlet for LES

In LES of compressible flows, acoustic waves can propagate through the numerical domain. If the discretization and the numerical method are chosen adequately, the acoustic waves are not significantly damped. In turbulent flows, the turbulence itself can generate noise. In combustion systems the flame can also act as a source of noise. This noise radiates away from the source, dissipating the acoustic energy through the domain. The boundary conditions typically used in CFD simulations (fixed value and fixed gradient) have the property of reflecting acoustic waves. Thus, acoustic energy cannot leave the simulation domain and accumulates in a non-physical manner. This issue can be addressed by using non-reflecting boundary conditions.

The non-reflective pressure boundary condition implemented in OpenFOAM is a simplification of that proposed by Poinsot and Lele [92] and is described by Kärrholm [55]. The model needs two constants, p_∞ and l_∞, representing the pressure at infinity and a relaxation length scale, respectively. The relaxation length scale is the parameter that governs how reflective the outlet will be; a low value will give a more reflective outlet than a high value. The model begins by calculating the velocity of the outgoing pressure wave ω:

$$\omega = U_i \cdot \vec{n} + \sqrt{1/\psi} \tag{5.128}$$

where $\vec{n}$ is the normal vector at the outlet and ψ is the compressibility of the media. The pressure wave velocity is used to calculate the pressure wave coefficient α and the relaxation coefficient b:

$$\alpha = \omega \frac{\Delta t}{\delta} \tag{5.129}$$

$$b = \omega \frac{\Delta t}{l_\infty} \tag{5.130}$$

where δ is the cell-face distance coefficient. The value of the pressure is calculated using these properties:

$$p_{trans} = \frac{p_0 + b p_\infty}{1+b} \tag{5.131}$$

$$\xi = \frac{1+b}{1+\alpha+b} \tag{5.132}$$

where p_0 refers to the pressure at the outlet in the previous time step. The pressure at the outlet is not set directly to p_{trans}, instead it is relaxed using the pressure in the cell p_{cell} closest to the outlet

$$p_{outlet} = \xi p_{trans} + (1 - \xi) p_{cell}. \tag{5.133}$$

where ξ is the relaxation factor.

5.7.4 Domain decomposition and parallel computing

Despite the even increasing performance of the computers available for CFD simulations, even a modest LES can take months to complete on a single computer. To shorten this time, the numerical domain is divided into a collection of subdomains, which are solved simultaneously in several computers connected in parallel.

The METIS algorithm [49] was used as decomposition strategy. METIS uses graph partitioning techniques to split the computational domain into subdomains, firstly converting the finite volume mesh into a graph, and then partitioning it in an optimal way. The partitioning ensures that the number of elements assigned to each processor is roughly the same and that the number of elements in the interface between different processors is minimized. The goal of the first condition is to achieve efficient load balancing for the computations among the processors. The second condition ensures that the required information exchange between different processors is minimized.

The information exchange between the partitions uses the software library OpenMPI [29] as message passing interface (MPI). MPI is an industry standard, typically used for parallel and distributed computing. OpenMPI is an open source, freely available implementation of the MPI standard.

6 Free Jet

Jets are the most commonly studied turbulent free shear flows, along with wakes and mixing layers. The name *free* implies that these flows are remote from walls, with the turbulence originating from differences of the mean velocity field. Even if the jet seems as a simplified flow configuration with limited practical importance, its analysis is a first step towards understanding and predicting more complex flows in which recirculation and curvature effects are present. In particular, the free jet has been used to test and validate the different numerical tools and the different simulation methodologies applied for the jet in crossflow, which will be described in the next chapter.

This chapter begins with a short review of different experimental setups found in the literature about the free jet. The text continues with the numerical setup used for the simulations, and finally the results of the simulations are compared with experimental data and discussed.

6.1 Experimental setup

The free jet has been already introduced in Chapters 3 and 4 (see Figures 3.3, 3.4 and 4.1).

Axisymmetric jets have been studied since the 1940s, with Corrsin [17] being responsible for one of the pioneer works in this field. Several works involving experimental data, mathematical analysis and computational modeling have followed, as the reviews by Hinze [39] and Rodi [101]. A modern reference about the theory of the turbulent axisymmetric jet is the textbook of Pope [93].

There are two main configurations used in experimental studies, which differ in the way the jet is produced. Smoothly contracting nozzles produce a plug-flow velocity profile with low turbulence level at the nozzle exit, in contrast to long straight pipes, which produce a fully developed, turbulent pipe flow. Experiments using the smoothly contracting nozzle have been selected in this work. The experimental configuration and the coordinate system are shown schematically in Figure 6.1. A Newtonian fluid steadily flows through a nozzle of diameter D, producing an approximately plug-flow velocity profile with mean velocity U_{jet}. The jet from the nozzle flows into a quiescent ambient of the same fluid. In the ideal case, the flow is completely defined by D, U_{jet} and the kinematic viscosity ν, and the only non-dimensional parameter is the Reynolds number defined by Re=$U_{jet}D/\nu$. However, in real configurations the details of the flow at the nozzle and the surroundings can affect the development of the jet [44].

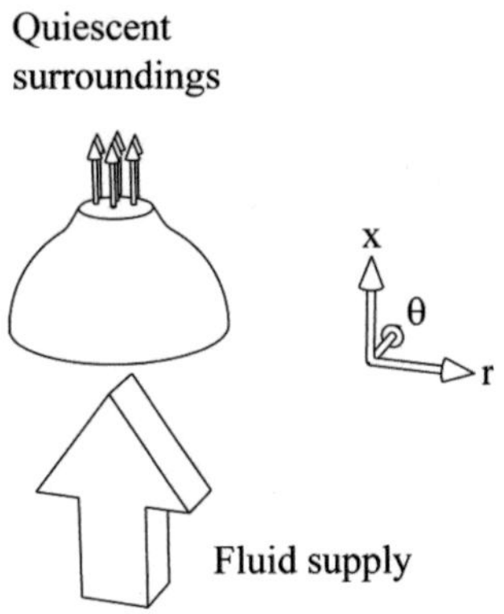

Figure 6.1: Sketch of the experimental configuration with a smoothly contracting nozzle and the cylindrical coordinate system employed.

The free jet is considered fully turbulent if the Reynolds number exceeds 10^4 [93]. The experiments of Panchapakesan and Lumley [86] and Hussein *et al.* [44] fit well in this region, using air as flow medium and with Reynolds numbers of 11 000 and 95 500, respectively. The meticulous work of Hussein *et al.* has employed three different methods for the measurements: stationary hot-wire anemometry, flying hot-wire anemometry and laser-Doppler anemometry (LDA). Their discussion about the limits and the validity of hot-wire anemometry data is especially interesting. The LDA results, being potentially more reliable, will be used as a benchmark in the comparison with the present simulations.

Experimental data about the mixing of a passive scalar is scarcer in the literature than velocity data. Mi *et al.* [75] have summarized some of the most important experimental investigations of the scalar field of axisymmetric turbulent jets. The experimental data of Dowling and Dimotakis [24] follow closely the numerical setup proposed in this work, with a Reynolds number of 16 000, and will be used as a benchmark. The measurements were made using a non-intrusive laser-Rayleigh scattering technique.

6.2 Numerical setup

As the results will be compared to various experiments, the computational domain was designed as general as possible. Figure 6.2 shows a sketch of the computational domain and the cylindrical coordinate system along with their respective velocity components. The jet discharges from a pipe with 10 mm diameter and 30 mm length into a domain with 800 mm diameter and 1000 mm length. The computational grid consists of 3.6 million hexahedral elements, with the jet inlet having a resolution of 32 grid elements along its diameter. The central part of the grid is depicted in Figure 6.2. A coarser grid consisting of 500 000 elements was also employed for some tests of the LES and for the RANS simulations.

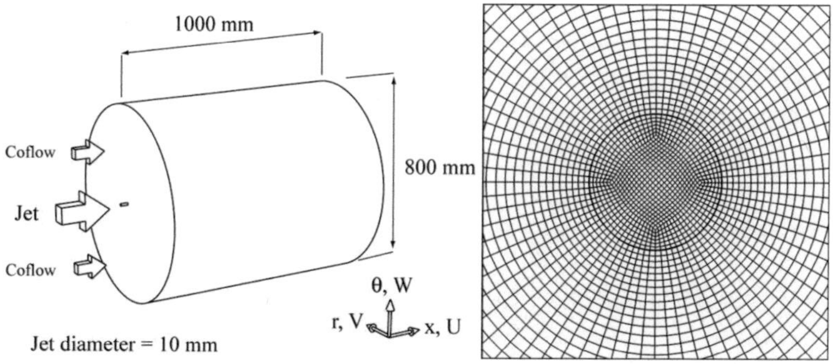

Figure 6.2: Overview of the computational domain with dimensions, left side. The central part of the grid in a plane perpendicular to the symmetry axis, right side. The thick circle line represents the jet inlet.

The jet inlet velocity was set to 100 m/s, resulting in a Reynolds number based on the pipe diameter and the mean velocity of about 73 000. In order to preserve the plug-flow velocity profile imposed at the jet inlet, the pipe walls were modeled as frictionless, which prevented the development of a boundary layer. The turbulence intensity was set to 1%, following approximately the value from Hussein *et al.* [44]. For the LES, the synthetic turbulence generated by the boundary condition described in Section 5.7.3 was employed. The values of k and ε or ω in the RANS simulations were calculated to represent the 1% turbulence intensity and a turbulent length scale of $\ell_t = 1/7D$.

Instead of a pure quiescent surrounding, the jet was introduced into a very weak coflow imposed at the bottom side of the domain, with 0.1 m/s mean velocity, without fluctuations. This strategy was employed in order to improve the stability of the simulations, especially the LES, as a pure free-stream boundary condition near the jet inlet affected negatively the numerical stability. Free-stream boundary conditions were used in the lateral and top boundaries.

For the LES, the time step was 1 microsecond long for the coarse grid and 0.5 microsecond long for the fine grid, yielding a maximum CFL number of approximately 0.3. A total of 0.9 seconds was simulated, resulting in 5 residence times calculated using the mean axial velocity at the end the computational domain and the domain length

$$\tau = \frac{\ell}{U} = \frac{1000mm}{6m/s} = 0.1667s. \tag{6.1}$$

The simulation time corresponds to 900 000 time steps for the coarse grid and 1.8 million time steps for the fine grid. The LES with the fine grid used for 28 days 128 processors of the HP XC4000 supercomputer of the Steinbuch Centre for Computing of the Karlsruhe Institute of Technology (KIT).

Turbulence modeling

Two subgrid scale turbulence models were available for the LES: the standard Smagorinsky model with fixed model parameters, and a dynamic version in which the model parameters are calculated from the smallest scales (see Section 5.4).

For the RANS simulations, two different models were used, the standard k-ε and the SST models (see Section 5.2). Unsteady RANS simulations have been attempted; however, the flow remained in steady state with virtually the same results as the steady state RANS simulations, so only the results of the latter simulations are presented in the following sections.

6.3 Results and discussion

6.3.1 Preliminary LES tests

The performance of two subgrid scale turbulence models of the LES, namely the standard and dynamic Smagorinsky models, was assessed using the coarse grid in order to select the most appropriate model for the LES using the fine grid.

For a self-similar jet, the mean centerline velocity $U_c(x)$ is given by

$$\frac{U_c(x)}{U_{jet}} = \frac{B_u}{(x-x_0)/D}, \tag{6.2}$$

where U_{jet} is the mean jet axial velocity at the inlet, B_u is an empirical constant, x_0 represents a virtual origin and D is the pipe diameter [93].

Figure 6.3 shows the inverse of the centerline velocity, specifically U_{jet}/U_c, plotted against $(x-x_0)/D$. The points represent the result of Equation (6.2) using B_u = 5.8 and x_0 = 4.0 D, as suggested by Hussein *et al.* [44]. The LES using the standard Smagorinsky model agrees better with the measurements of Hussein *et al.* than the dynamic version. The meaning of the curve slope will be discussed in more detail using the results of the Figure 6.7; it suffices here to state that the smaller curve slope of the dynamic version means that the jet is less diffusive than the measurements.

Another characteristic variable of the free jet is the half-width $r_{1/2}$, which is defined as the radial position where the mean axial velocity is one-half of its centerline value. Employing the half-width $r_{1/2}$, the spreading rate S can be calculated as

$$S = \frac{dr_{1/2}(x)}{dr}. \tag{6.3}$$

In the self-similar region of turbulent jets, the jet spreads linearly and S is a constant, and an empirical law for $r_{1/2}$ can be derived

$$r_{1/2}(x) = S(x-x_0). \tag{6.4}$$

The cross-stream similarity variable in free jets can be taken to be either $r/r_{1/2}$ or $r/(x-x_0)$. Using the facts that S is a constant $S = 0.0094$ and that the virtual origin $x_0 = 0$, the two variables are related by $r/x- = S(r/r_{1/2})$.

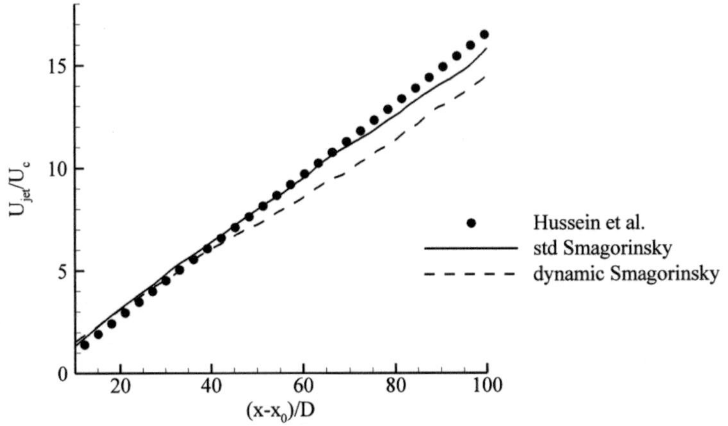

Figure 6.3: Line plots of the inverse of the centerline velocity U_{jet}/U_c with distance from jet from the LES using the standard and dynamic versions of the Smagorinsky subgrid scale turbulence model, with the coarse grid.

Figure 6.4 shows the radial velocity profiles using both cross-stream similarity variables. The profiles have the same behavior seen in Figure 6.3, where the LES with the dynamic version is less diffusive than both measurements and the LES with the standard Smagorinsky. As the standard version of the Smagorinsky model showed a better performance, it was used in the simulation using the fine grid presented in the next sections.

The value of the turbulent Schmidt number σ_t used to model the subgrid scale flux vector (see Section 5.4) varies significantly in the literature. Values of σ_t from 0.7 [8, 21], to 0.9 [23] and 1.0 [103, 124] have been proposed. The influence of σ_t has been assessed using simulations with the coarse grid and the standard Smagorinsky model, comparing the mean passive scalar when using σ_t = 1.0 and 0.7.

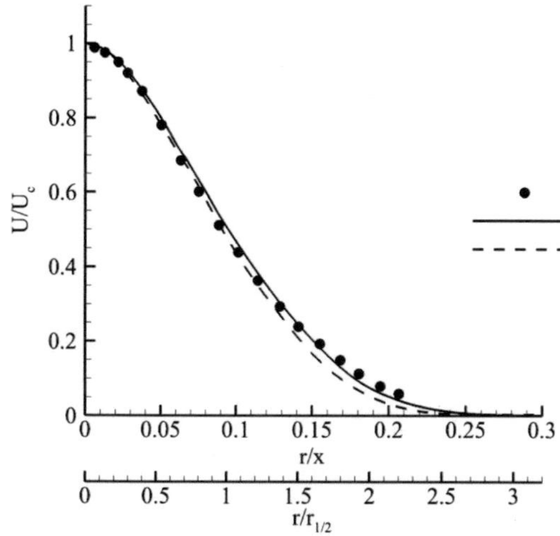

Figure 6.4: Radial profiles of axial velocity normalized by the centerline velocity U/U_c from the LES using the standard and dynamic versions of the Smagorinsky subgrid scale turbulence model, with the coarse grid.

For the passive scalar, the mean centerline value $C_c(x)$ is given by

$$\frac{C_c(x)}{C_{jet}} = \frac{B_c}{(x - x_0)/D}, \tag{6.5}$$

where C_{jet} is the value of the mean passive scalar at the inlet and B_c is an empirical constant, different from the constant B_u used for the centerline velocity [93].

The quantity C_{jet}/C_c is plotted versus $(x-x_0)/D$ in Figure 6.5. Equation (6.5) is represented by the points, using the constants $B_c = 4.73$ and $x_0 = 0.0$ as suggested by Dowling and Dimotakis [24]. The agreement of the simulations with Equation (6.5) is worse than the agreement of the mean velocity, which is probably a consequence of the grid being too coarse. The agreement is, however, good enough for the parameter variation study being proposed. The difference between the two simulations is very small, which indicates a limited influence of the subgrid scale on the mean values.

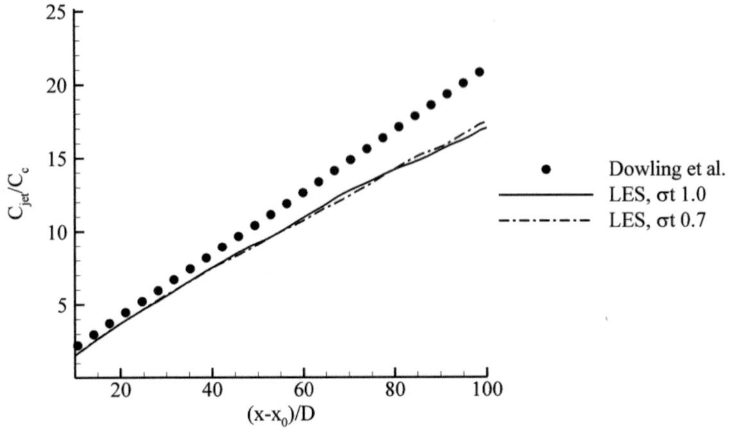

Figure 6.5: Line plots of the inverse of the passive scalar at the centerline C_{jet}/C_c with distance from jet from the LES using the standard Smagorinsky model and turbulent Schmidt numbers σ_t of 0.7 and 1.0, with the coarse grid.

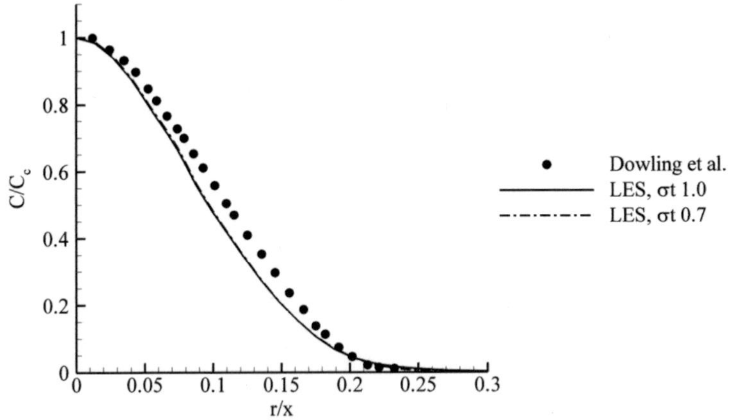

Figure 6.6: Radial profiles of the mean passive scalar normalized by its centerline value C/C_c from the LES using the standard Smagorinsky model and turbulent Schmidt numbers σ_t of 0.7 and 1.0, with the coarse grid.

Figure 6.6 shows the mean passive scalar normalized by its centerline value, C/C_c, along r/x. The points represent the measurements of Dowling and Dimotakis [24], and the agreement is better than with the axial values. The difference between the two simulations is again very small.

As the simulations have shown almost no sensitivity to the value of σ_t, the value of 1.0 has been used in the following simulations.

6.3.2 Axial profiles

Figure 6.7 shows the centerline velocity of the LES using the standard Smagorinsky model and the fine grid, and the RANS simulations with the k-ε and the SST models. The points again represent the result of Equation (6.2) using $B_u = 5.8$ and $x_0 = 4.0$ D. The virtual origin x_0 is a geometry-dependent value, and its value in the simulations is approximately zero. The expected $1/x$ decay rate is represented as a straight line on the plot. The greater the slope of the line, the smaller the value of B and the higher the decay rate of the centerline velocity. The curve slope of the LES is slightly smaller than the experiments. The slope using the k-ε model is slightly greater than the measurements, however still inside the scatter of experimental values. The simulation with the SST turbulence model has an even greater slope, resulting in a too diffusive jet.

Figure 6.8 shows C_{jet}/C_c plotted versus $(x - x_0)/D$. Equation (6.5) is represented by the points, using the constants $B_c = 4.73$ and $x_0 = 0.0$ as suggested by Dowling and Dimotakis [24]. The LES shows a slightly smaller slope compared to the measurements, indicating that not enough diffusion has been modeled. The RANS simulations of the passive scalar have an extra parameter, the turbulent Schmidt number σ_t, which is the quotient between the turbulent viscosity and diffusivity (see Equation (5.30)). The smaller the σ_t, the greater the Reynolds flux and consequently the turbulent diffusion. The recommended value for this constant varies between 0.7 for free shear flows and 0.9 for boundary layers. Using the k-ε model and $\sigma_t = 0.7$, the centerline value of the passive scalar has a very good agreement with the measurement. When using $\sigma_t = 0.9$, however, the curve slope is smaller than the measurement, which means that not enough diffusion was present. The tendency of smaller diffusive transport with increasing value of σ_t is the same using the SST model, but the slope using $\sigma_t = 0.7$ is too great, and using $\sigma_t = 0.9$ it compares well with the measurement.

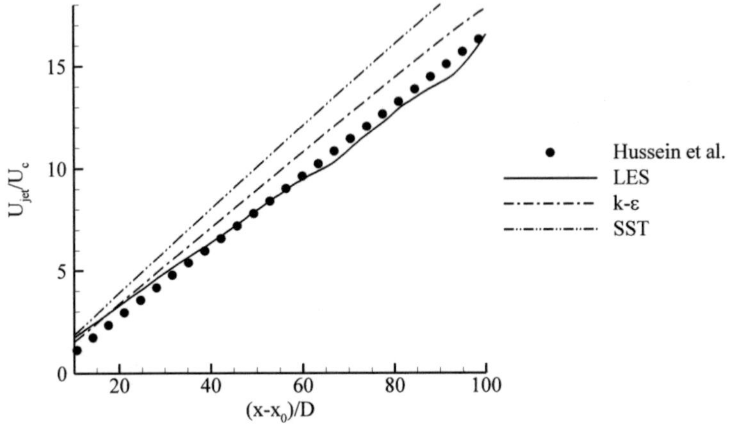

Figure 6.7: Line plots of the inverse of the centerline velocity U_{jet}/U_c with distance from jet.

Figure 6.9 shows the fluctuation of the axial velocity component $\overline{u'}$ along the jet axis. The quantity $\overline{u'}/U_c$ in the LES tends asymptotically to a value close to 0.24 after the development region, which agrees well with the measurements of Panchapakesan and Lumley [86] and Hussein *et al.* [44], which indicate values of 0.24 and 0.27, respectively. In order to show the amount of the axial velocity fluctuation being resolved by the large-scale structures of the LES, $\overline{u'}/U_c$ without the subgrid scale (SGS) fraction is also plotted in Figure 6.9. The modeled fraction of $\overline{u'}$ is of the order of 3% of the total value, which is an indication of the good resolution of the LES. The RANS simulations show a slightly more intense turbulence at the axis, with asymptotic values of 0.27 and 0.295 using the k-ε and SST models, respectively.

The normalized root-mean-square (rms) of the passive scalar fluctuation c_{rms}/C_c along the centerline is shown in Figure 6.10. Only the result of the LES is shown, as the RANS simulations do not evaluate c_{rms} directly; a transport equation needs to the resolved if the value of c_{rms} is of relevance. Interestingly, the normalized fluctuation c_{rms}/C_c reported by Dowling and Dimotakis [24] tends asymptotically to a value between 0.23 and 0.24, which is close to the value of 0.25 for the velocity fluctuation. The result of the LES tends to a smaller value around 0.21, *i.e.*, the diffusion level in the simulation is lower than in the measurements.

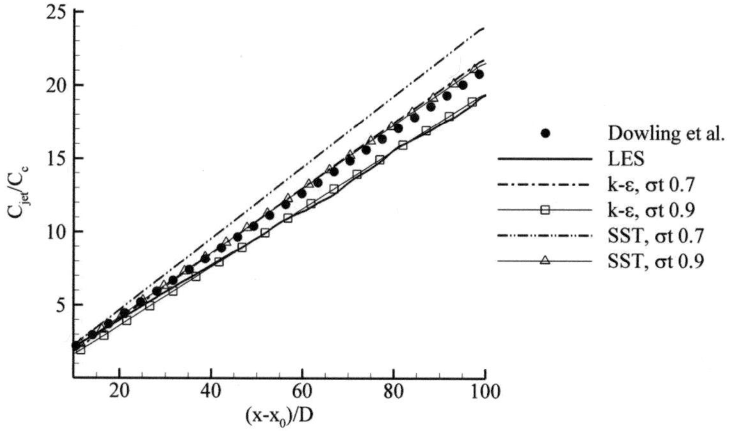

Figure 6.8: Line plots of the inverse of the passive scalar at the centerline C_{jet}/C_c with distance from jet.

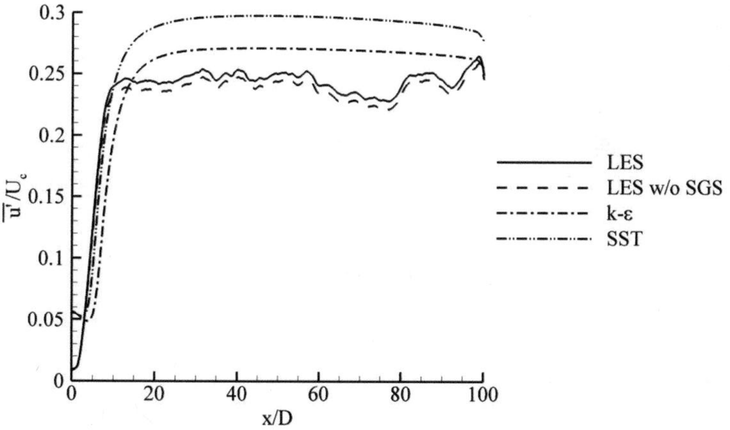

Figure 6.9: Line plots of axial velocity fluctuation $\overline{u'}/U_c$ with distance from jet.

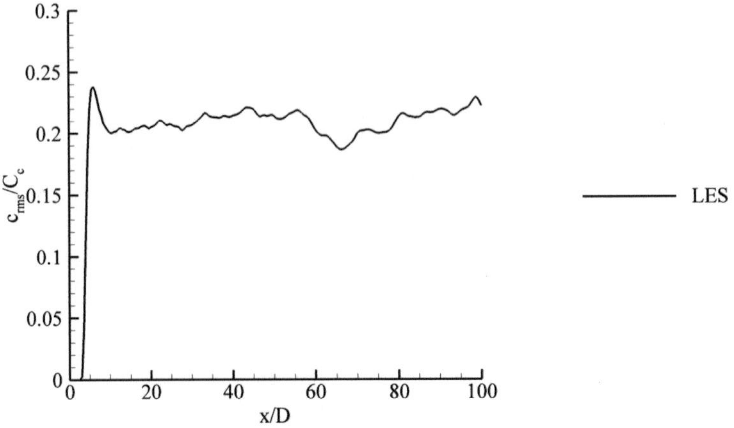

Figure 6.10: Line plots of passive scalar rms c_{rms}/C_c with distance from jet.

6.3.3 Radial profiles

Plotted against the cross-stream similarity variable r/x, the normalized radial mean velocity U/U_c profiles of the LES for different axial positions approximately collapse, as can be seen in Figure 6.11, indicating that the profiles are self-similar. The measurements of Hussein *et al.* [44] also collapse after a distance of 30 diameters, reaching the self-similar region. The normalized velocity profiles U/U_c of the measurements and the simulations are plotted in Figure 6.12, where the LES velocity profile was averaged over the various axial positions seen in Figure 6.11. As in the case of the velocity along the axis, the radial profiles of the LES are in very good agreement with the measurements. The profiles of the RANS simulation are broader, an indication of too much diffusion, with the SST model being more diffusive than the k-ε model. The jet half-width $(r/x)_{1/2}$ has the value 0.094 in the measurements. The LES has approximately the same half-width as the measurements, while the values of the RANS simulations are larger, with values of 0.103 and 0.119 for the k-ε and SST models, respectively.

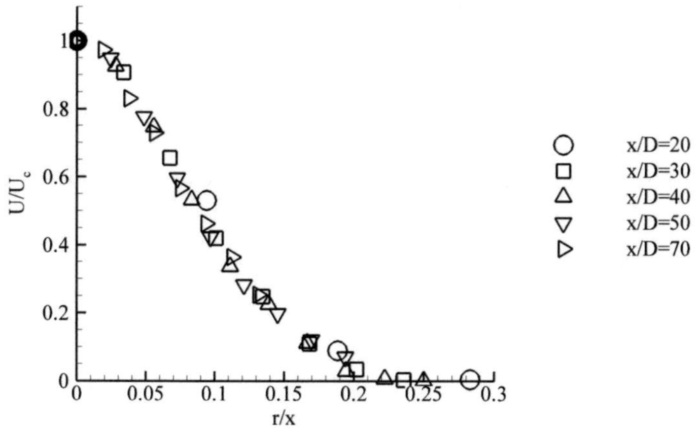

Figure 6.11: Radial profiles of axial velocity normalized by the centerline velocity U/U_c for various axial positions of the LES.

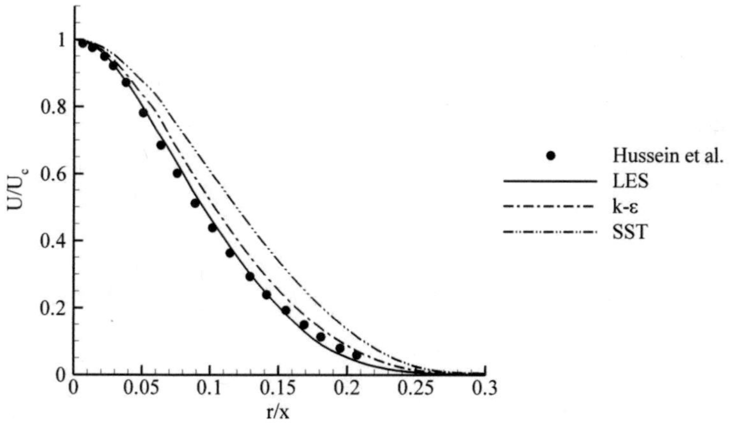

Figure 6.12: Radial profiles of axial velocity normalized by the centerline velocity U/U_c.

The mean passive scalar normalized by its centerline value, C/C_c, along r/x are shown in Figure 6.13. The points represent the measurements of Dowling and Dimotakis [24]. The measurements show a half-radius value of 0.114. The agreement of the LES results with the measurements is very good. The RANS simulation with k-ε shows good agreement using $\sigma_t = 0.9$, but is too diffusive with $\sigma_t = 0.7$, with half-radius values of 0.110 and 0.126, respectively. The simulations with the SST model are too diffusive using both values of σ_t, with half-radius values of 0.145 for $\sigma_t = 0.7$ and 0.127 for $\sigma_t = 0.9$.

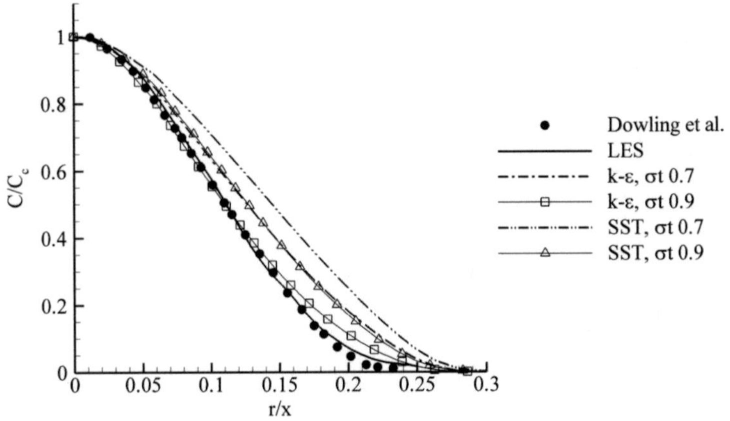

Figure 6.13: Radial profiles of the mean passive scalar normalized by its centerline value C/C_c.

The evolution of the specific Reynolds stress components $\overline{u'u'}$, $\overline{v'v'}$, $\overline{w'w'}$ and $\overline{u'v'}$ non-dimensionalized by the square of the centerline velocity can be seen in Figure 6.14. To improve the readability of the diagram, only the LES and the RANS simulation with k-ε are shown. The Reynolds stress components of the steady state RANS simulations have been calculated using the Boussinesq hypothesis [121]. The $\overline{u'u'}/U_c^2$ is the most intense component, with a value at the axis of 0.075 in the measurements of Hussein *et al.* [44]. The axisymmetry of the flow requires $\overline{v'v'}$ and $\overline{w'w'}$ to be equal, a condition nearly satisfied by the measurements and by the simulations. The shear stress $\overline{u'v'}$ is the only component that is zero at the axis. While $\overline{u'v'}$ is predicted well by both simulations, the normal stresses show more discrepancies. The LES is able to reproduce the large difference between $\overline{u'u'}$ and the other normal stresses near the axis, as shown by the measurements. The k-ε model is too simplistic to fully capture this effect, and the levels remain very close to each other.

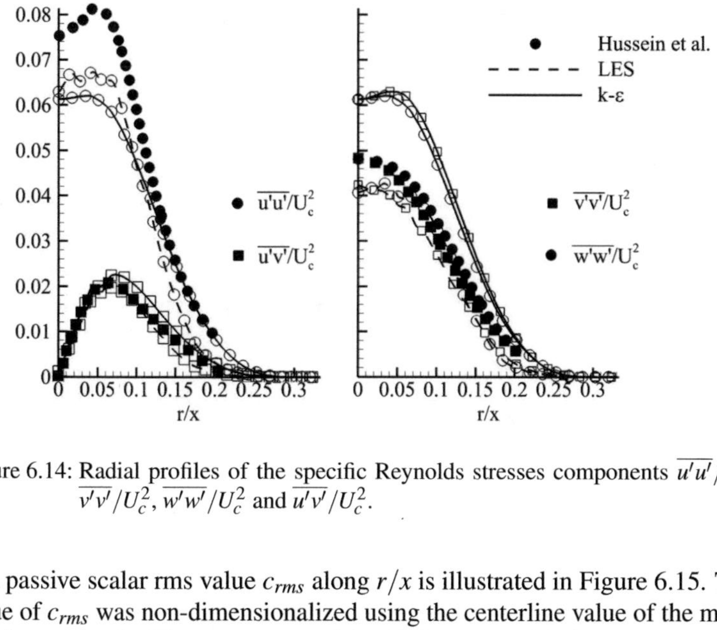

Figure 6.14: Radial profiles of the specific Reynolds stresses components $\overline{u'u'}/U_c^2$, $\overline{v'v'}/U_c^2$, $\overline{w'w'}/U_c^2$ and $\overline{u'v'}/U_c^2$.

The passive scalar rms value c_{rms} along r/x is illustrated in Figure 6.15. The value of c_{rms} was non-dimensionalized using the centerline value of the mean passive scalar C_c. The agreement of the LES results with the measurements is good; however, the overall fluctuation level is underpredicted. The RANS simulations do not evaluate c_{rms} directly; a transport equation needs to be solved if the value of c_{rms} is of relevance. In the present RANS simulations this equation has not been solved, so no results for c_{rms} are shown.

6.3.4 Correlation between momentum and passive scalar turbulent diffusivity

The correlation between momentum and passive scalar turbulent diffusivity within the RANS context is modeled by the turbulent Schmidt number σ_t

introduced in Section 5.2. The same information can be determined from the LES and compared to the value used in the RANS simulations. In the case of the free jet, the mean passive scalar flux is effective only in the radial direction. This fact significantly simplifies the analysis, as only one pair of Reynolds stress and Reynolds flux components has to be taken into account. The same analysis for the jet in crossflow configuration is more complex, as will be discussed in the next chapter. The turbulent Schmidt number from the LES can be calculated using the following equation

$$\overline{u'v'}/\overline{u'c'} = \sigma_t \frac{2S_{ij}}{\partial C/\partial y}. \tag{6.6}$$

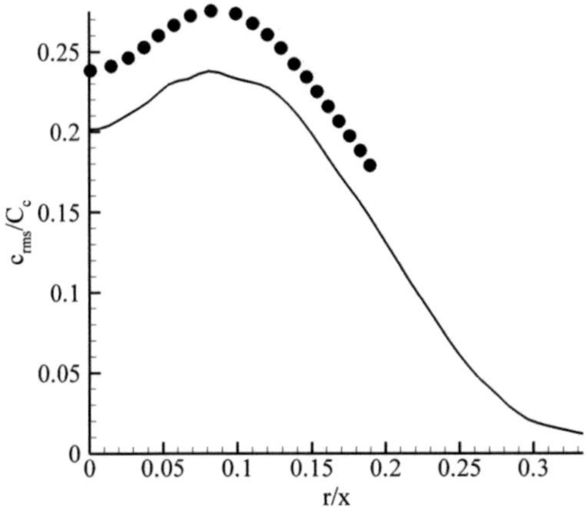

Figure 6.15: Radial profiles of passive scalar fluctuation c_{rms}/C_c.

Figure 6.16 shows the radial profile of σ_t obtained from the LES. It should be noticed that the turbulent Schmidt number is difficult to extract from the simulations, as it is derived from mean fluctuation values and gradients of the mean variables, both very sensitive to the run time of the LES. As the averaging process is more effective in regions with higher velocity, the mean values are smoother in the region near the jet inlet, especially the mean fluctuation values $\overline{u'v'}$ and $\overline{u'c'}$. The profiles of σ_t could be calculated with good accuracy up to a distance x/D=30. Another point that should be noticed is that at the jet axis the gradients of the mean values tend to zero, which leads to a singularity at the axis; the values of σ_t at these positions were blanked. The mean value of σ_t is 0.75, which compares well with the suggested value of 0.72 from Hinze [39] and the value of 0.7 from the experimental observations of Yimer *et al.* [123]. The good agreement is a further indication of the high quality of the LES results.

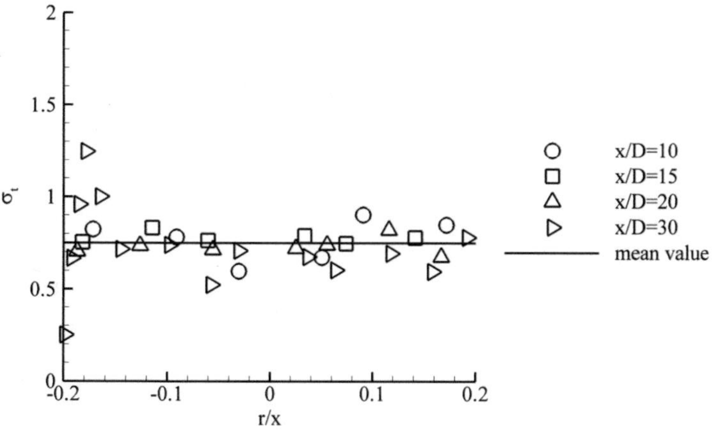

Figure 6.16: Radial profile of σ_t for the LES.

6.3.5 Coherent structures and intermittency

The present LES data allow the investigation of the formation and evolution of coherent structures. For this purpose, the isosurface of the pressure fluctuation $p - \overline{p}$ is employed in Figure 6.17. The constant value for obtaining the isosurface is close to zero, and was chosen to enhance the presentation of the structures. Shortly downstream of the jet inlet, vortex rings due to the Kelvin-Helmholtz instability can be observed. Although the visualization technique differs, there is a similarity between these structures and the Kelvin-Helmholtz instabilities in the LIF cross section shown in Figure 3.3. In addition, the finding of large-scale organization of entrainment and mixing in the self-similar field of the jets by Dahm and Dimotakis [19], which resulted in the conceptual picture in Figure 3.4, correlate well with the current results.

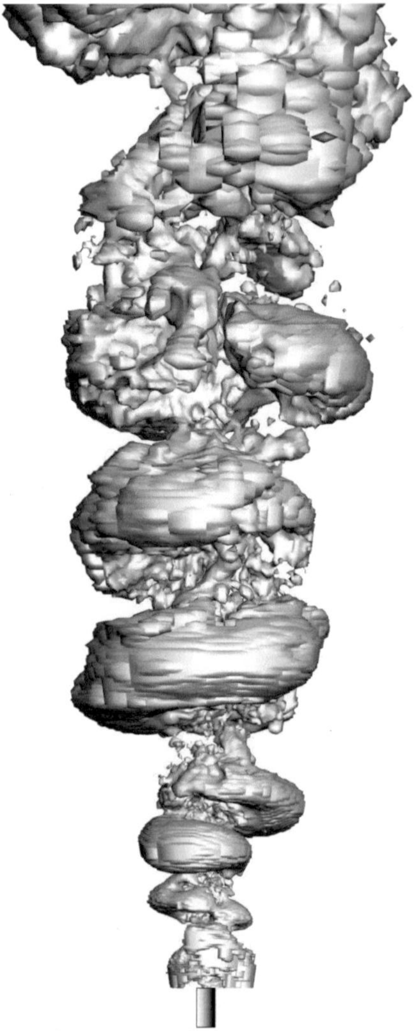

Figure 6.17: Isosurface of pressure fluctuation.

The large-scale structures seen in the LES form the regions with fairly uniform composition proposed by Dahm and Dimotakis and depicted in Figure 3.4. Further downstream the vortex rings can no longer be discerned, giving place to larger coherent structures.

The flow is intermittent, as can be expected from the very evident coherent structures. The intermittency function, Equation (4.1), was used to calculate conditional averages. The mean values conditioned with the turbulent jet flow have been assigned with the subscript t, while the values conditioned with the ambient fluid flow have been assigned with the subscript n. The threshold value of 1% of the passive scalar has been used for calculating the intermittency function, as suggested by Libby [63]. The radial evolution of the intermittency factor γ along with the mean, turbulent jet mean and ambient fluid mean values of the axial velocity and passive scalar is shown in Figures 6.18 and 6.19, respectively. The measurements of Antonia *et al.* [2] for a turbulent heated jet in a co-flowing stream, presented in Chapter 4, show a similar trend although the setups are quite different.

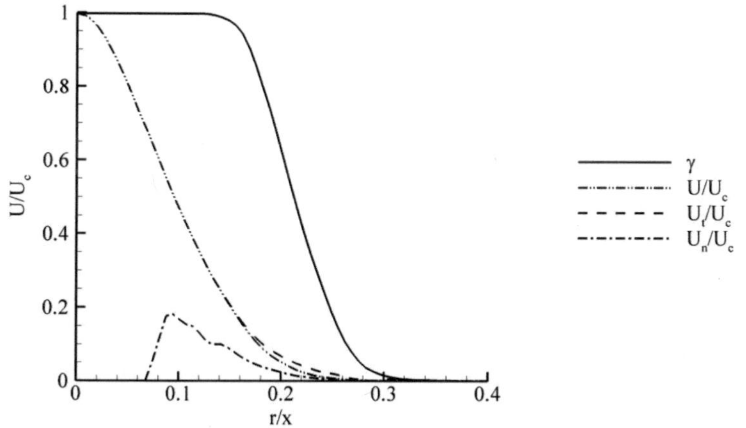

Figure 6.18: Radial profiles of intermittency factor γ, mean axial velocity U/U_c, turbulent mean axial velocity U_t/U_c and non-turbulent mean axial velocity U_n/U_c for the LES.

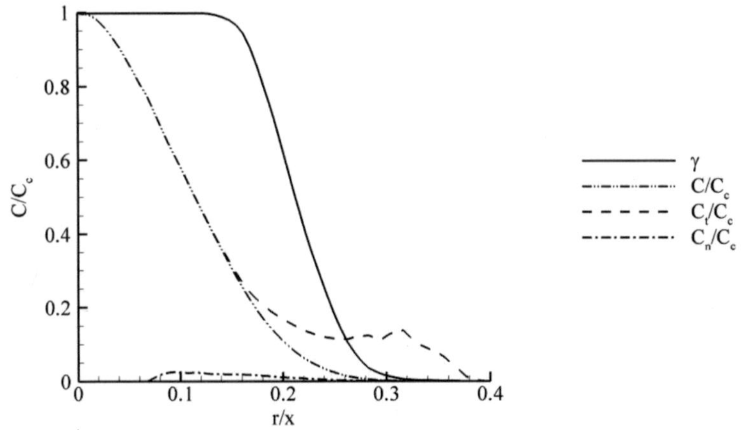

Figure 6.19: Radial profiles of intermittency factor γ, mean passive scalar C/C_c, turbulent mean passive scalar C_t/C_c and non-turbulent mean passive scalar C_n/C_c for the LES.

The analysis of the γ profiles shows that the flow is always turbulent near the jet axis until $r/x \approx 0.15$, where the function has the value unity. Between this point and the jet edge, the fraction of the time that each position remains in the turbulent flow decreases until reaching zero, followed by the γ function. The influence of the intermittency is restricted to this region.

The mean velocity profiles are shown in Figure 6.18. The difference between the mean value U and the jet fluid mean value U_t of the velocity is not very evident, even in the highly intermittent area. On the other hand, the mean velocity U_n of the ambient fluid found in the intermittent region has higher values than the quiescent surrounding, approaching 10% of the centerline value.

The profiles of the passive scalar in Figure 6.19 show a different behavior than the velocity profiles, revealing some interesting features of the jet. The value of the mean passive scalar C_t of the jet fluid is considerably higher than the mean value C in the intermittent region. It indicates that the mean value diminishes when approaching the jet edge as a consequence of the diminishing fraction of time that the jet fluid spends in this region, however the passive scalar value inside the jet remains more constant toward the jet edge.

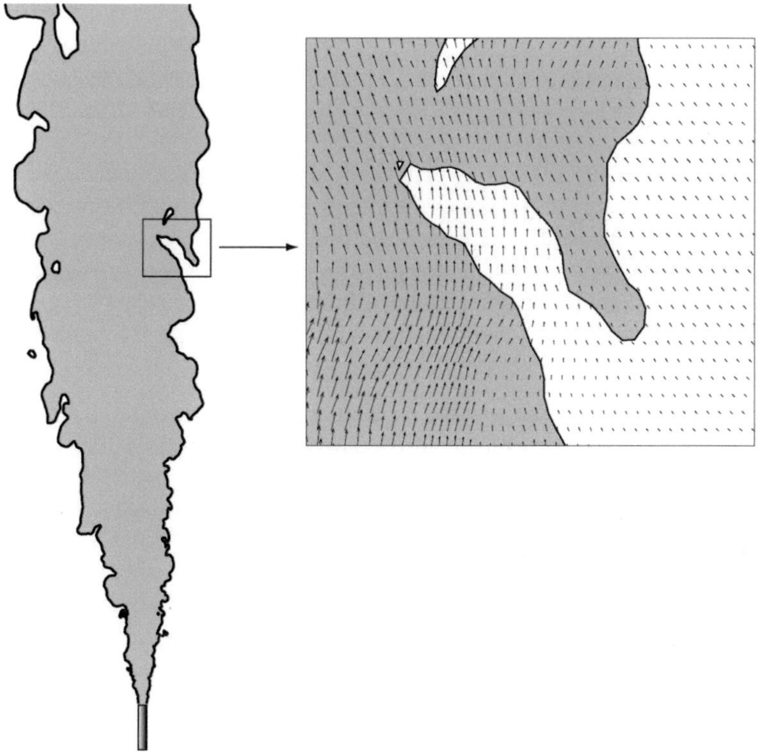

Figure 6.20: Symmetry plane of one realization of the LES. The solid line represents the boundary between jet flow and the ambient fluid. The detail shows also arrows representing the local velocity vector.

The different behavior of the turbulent mean velocity in comparison to the turbulent mean passive scalar is attributed to velocity fluctuations within the ambient fluid entrained by the jet flow. These fluctuations, that are not turbulent in nature, are induced by the pressure fluctuations associated with the turbulence [63]. This effect is illustrated in Figure 6.20, in which the symmetry plane of the LES is depicted. The boundary between the jet flow and the ambient fluid is marked by an isocontour of 1% passive scalar. As expected, this boundary is highly convoluted, a result of the high intermittency. A small region where entrainment of ambient fluid is taking place is shown in more detail, with arrows representing the local velocity vector. It clearly shows that the ambient fluid in the entrained region has higher velocity there than at the surroundings. This finding corroborates with the theory [1, 93] and the experiments of Antonia *et al.* [2] in Figure 4.6, which dictates that the turbulent jet fluid has a higher streamwise velocity than both the ambient fluid that is entrained and the velocity of the surrounding fluid. Ambient fluid is entrained mainly at locations involving significant intermittency, and its velocity is higher than the value of the surrounding fluid. It follows that the velocity variation across the superlayer is smaller than the variation of the passive scalar, which accounts for the apparent greater width of the scalar profile relative to the velocity profile. In Figure 6.20, for example, the non-turbulent mean velocity of the fluid at locations with high intermittency is higher than the value of the ambient fluid.

Analyzing the diffusion across the superlayer helps to understand the discrepancy between the velocity and passive scalar fields. In the classic Fick's law of diffusion, the driving force of diffusion is the gradient of the variable being analyzed. To be able to compare the gradients directly, the u velocity component was divided by the mean value at the inlet, creating the dimensionless velocity u/U_{jet} that spans from zero to one as the passive scalar does. Following this idea, the diffusion flux of velocity J_u and passive scalar J_c across the superlayer can be defined as the dot product of the gradients and the normal vector $\vec{n}$

$$J_u = \nu \frac{\partial (u/U_{jet})}{\partial x_i} \cdot \vec{n} \tag{6.7}$$

$$J_c = D \frac{\partial c}{\partial x_i} \cdot \vec{n} \tag{6.8}$$

where ν is the kinematic viscosity, D is the scalar diffusivity, and $\vec{n}$ is the normal area vector of the superlayer. In these definitions, the subgrid scale viscosity ν_t and diffusivity D_t have been neglected.

To apply this formalism to the LES, the superlayer was approximated by the surface determined by the isosurface of 1% passive scalar, which is the same used to define the intermittency function. Integrating over the whole isosurface of a snapshot of the LES resulted in a dimensionless velocity flow rate of $8.24 \cdot 10^{-6}$ m^3/s and a passive scalar flow rate of $40.25 \cdot 10^{-6}$ m^3/s, which indicates that the passive scalar gradient is greater than the velocity gradient across the superlayer. In the LES, the turbulent Schmidt number is equal to one $\sigma_t = 1$, *i.e.*, the diffusivity of momentum and of the passive scalar are the same in the subgrid scale. The higher mean diffusivity of the scalar correlated with the higher diffusion flux J_c can only be explained by the effect of the large-scale structures.

The finding that the passive scalar flow rate across the superlayer is greater than the velocity flow rate also corroborates with the fact that the spreading rate of the passive scalar is larger than the spreading rate of the velocity.

6.4 Concluding thoughts

Despite of the complexity arising from the coherent structures of the flow, steady state RANS are able to predict important features of the free jet with good accuracy. The condition for a good simulation is the right choice of turbulence model and constants. The best overall results with RANS simulations have been obtained using the k-ε turbulence model and a turbulent Schmidt number σ_t of 0.7.

It was shown that the LES can reproduce the coherent structures of the jet flow, which translates into improved results in comparison to the RANS simulations. The LES results agree well with all measurements of the free jet. The anisotropy of the turbulence, for example, is good reproduced by the LES, while the RANS turbulence models are too simplistic in this particular area and cannot reproduce it in its full extension.

There is, however, a systematic discrepancy between the RANS results and the measurements in the high intermittent region, which is more evident in the profiles for passive scalar. Figure 6.21 shows the radial profiles of axial velocity and passive scalar side by side for the LES and the RANS simulation with k-ε model and $\sigma_t = 0.9$ and the measurements of Hussein *et al.* [44] and Dowling and Dimotakis [24]. While for the axial velocity the difference between the measurement and both simulations is small, the profiles of passive scalar show more discrepancy when approaching the jet edge. The scale of the diagram makes the difference appear small, but in fact it can be as high as 50% when approaching the jet edge. This discrepancy is related to the coherent structures and the intermittency, as their influence is different for the velocity and passive scalar fields. While the coherent structures and the intermittency are not resolved by the RANS simulations, leading to the observed discrepancy, the LES is able to reproduce them in this region, resulting in a better agreement with the measurements.

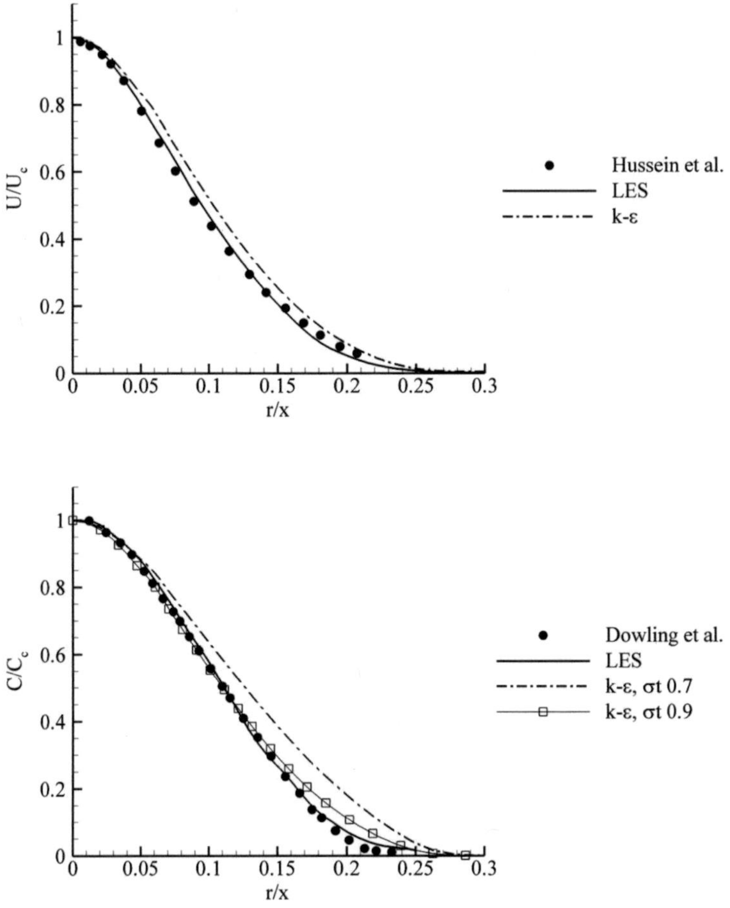

Figure 6.21: Radial profiles of axial velocity and passive scalar for simulations using LES and k-ε models.

7 Jet in Crossflow

The jet in crossflow (JIC) is a flow configuration of major theoretical and practical importance. It is more complex than the free jet, with a series of large-scale coherent structures and recirculation regions. This flow configuration is frequently found in many technical devices like mixers in the process industry, plumes from chimneys, engine exhaust gas pipes and cooling towers. Due to its good air-fuel mixing capability over a small distance, JIC is also favored by gas turbine manufacturers that use it in their lean premix burner technology, where the fuel and combustion air streams are premixed before the reaction zone.

This chapter begins with a review of the experimental investigations about the JIC. It is followed by the numerical setup used in the simulations. The chapter ends with the comparison of the results of the simulations with experimental data and the discussion of these results.

7.1 Experimental setup

In the discussion about coherent structures in Chapter 3, the various structures associated with the jet in crossflow (JIC) have been illustrated in Figure 3.5, which is plotted again in Figure 7.1 for convenience. The interaction between the two streams is associated with the formation of various turbulent vortices and complex coherent structures.

Numerous investigations of the JIC configuration have been performed especially focusing on the complex system of vortices and their contribution to the stability of the flow field. This is excellently summarized by Margason [72] who reviewed the investigations in the 1980s and at the beginning of the 1990s. Laser diagnostic tools have made tremendous progress in recent years, also benefiting studies about the JIC.

The turbulent mixing within the JIC under highly turbulent conditions has been investigated in the Engler-Bunte-Institute, Division of Combustion Technology

(EBI-VBT) by Georg Donnert, Camilo Cárdenas and Julia Sedlmaier, and the results are summarized on references [67, 30, 31, 18]. The experiments have been designed to provide unambiguous boundary conditions along with high quality local validation data.

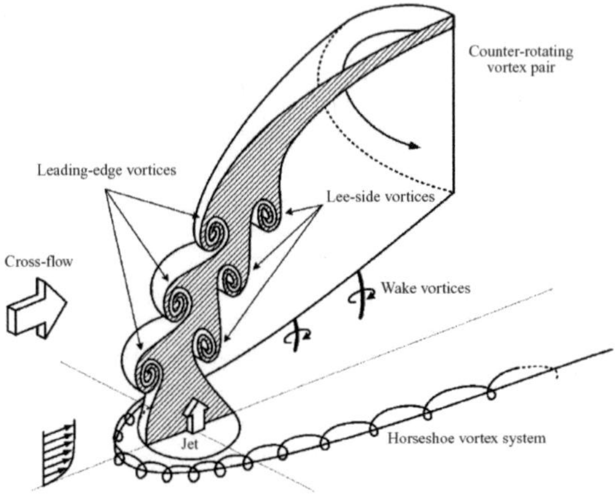

Figure 7.1: Scheme of the jet in crossflow phenomenology. Reproduced from [81].

In contrast to the literature data which employs mostly laminar or weakly turbulent flows to study the JIC, the experiments have been conducted under highly turbulent conditions with the crossflow Reynolds number spanning between $11.16 \cdot 10^4$ and $3.99 \cdot 10^4$ while the jet flow Reynolds number spans between $1.81 \cdot 10^4$ and $1.92 \cdot 10^4$, as can be seen in Table 7.1. The Reynolds numbers have been calculated using the channel cross section length and the jet inlet diameter as length parameters, respectively. The experimental facility is thoroughly described in references [14, 67]. It consists of a channel with square cross section (108 x 108 mm) in which a round jet (inner diameter D = 8 mm) is mounted flush to the wall. The pipe used to feed the jet is long enough to ensure a fully developed velocity profile. The center of the jet is placed 328 mm downstream of the beginning of the channel, where a plug flow velocity profile is generated by a specially built contraction nozzle. Optical access to the channel is given by four fused quartz windows placed at each side of the test section.

The main parameter that characterizes a jet in crossflow is the jet-to-crossflow velocity ratio, R. It is defined as the square root of the momentum ratio:

$R = (\rho_{jet} U_{jet}^2 / \rho_{cross} U_{cross}^2)^{1/2}$, where U_{jet} represents the bulk jet velocity and U_{cross} represents the bulk crossflow velocity. When the jet and crossflow densities are equal ($\rho_{jet} = \rho_{cross}$), the velocity ratio can be simplified to $R = U_{jet}/U_{cross}$. The experimental conditions for the four cases investigated in this study are described in Table 7.1. The turbulence intensity of the jet flow is 7% and of the crossflow is 1.5% for all cases.

		Case A	Case B	Case C	Case D
Jet	Bulk velocity		37.72 m/s		
	Re		$1.96 \cdot 10^4$		
Crossflow	Bulk velocity	9.08 m/s	15.95 m/s	8.40 m/s	5.70 m/s
	Re	$6.36 \cdot 10^4$	$11.16 \cdot 10^4$	$5.88 \cdot 10^4$	$3.99 \cdot 10^4$
	Velocity ratio	4.15	2.36	4.49	6.62

Table 7.1: Boundary conditions.

The measurement technique is illustrated in Figure 7.2. It consisted of a combination of two laser diagnostic methods: Particle Image Velocimetry (PIV) and Laser Induced Fluorescence (LIF). The main measurements have been conducted by Georg Donnert, with additional measurements by Camilo Cárdenas and Julia Sedlmaier; the results for Case A have been published by Galeazzo *et al.* [31], and combine PIV measurements at the symmetry plane with simultaneous PIV and LIF measurements at horizontal planes. For Cases B to D, simultaneous PIV and LIF measurements at the symmetry plane have been measured for different velocity ratios.

A frequency doubled Nd:YAG laser with an excited wavelength of 532 nm is used as light source. A Galilean telescope fans out the laser light beam, which is then guided it through the test section. This light sheet provides two different signals in the mixing region in the JIC arrangement. The scattered light (PIV) comes from aerosol particles added to both flows at a wavelength of 532 nm, while the fluorescence signal (LIF) is generated by excited NO_2 molecules added only to the jet flow at a range of wavelengths between 550 nm and 690 nm (Gulati et al. 1994). These two signals are spectrally separated and acquired by two different cameras. A CCD-camera (Dantec® 80C60 HiSense PIV- Camera, 1280 x 1024 pixels) is used for the scattered light while an intensified CCD-camera (Roper Scientific® 512 x 512 pixels) is used for the fluorescence light. A commercial program (Dantec FlowManager® Version 1.10) was used to post-process the PIV signals. The LIF signals were post-processed using an program developed in-house. Both signals were acquired simultaneously, allowing spatial measurements of the instantaneous velocity and concentration fields and their correlation.

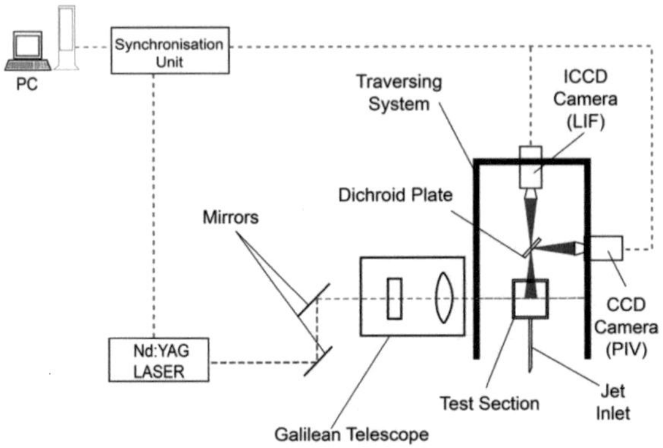

Figure 7.2: Overview of the 2D-PIV/LIF measurement technique.

An imaging area of 27.3 x 27.3 mm was used to completely resolve the flow phenomena including the high velocity gradients. An interrogation area of 32 x 32 pixels with an overlap of 50% was used for the PIV evaluation. The statistics were evaluated from a total of 6200 samples per measuring area.

7.2 Numerical setup

The computational domain was chosen in a way to save computational time, while capturing all important phenomena of the JIC. The coordinate system was centered at the center of the jet inlet into the crossflow channel. The domain for the large eddy simulation (LES) extended 100 mm in the upstream crossflow direction and 200 mm in the downstream direction, as depicted in Figure 7.3. A pipe with a diameter of 8 mm and a length of 40 mm was included in the domain.

Two grids have been used for the simulations, employing hexahedral-shaped elements. The coarse grid is composed of 1.5 million elements, with a resolution of 27 elements along the jet inlet diameter, while the fine grid is composed of 7.5 million elements, with an increased jet inlet resolution of 39 elements.

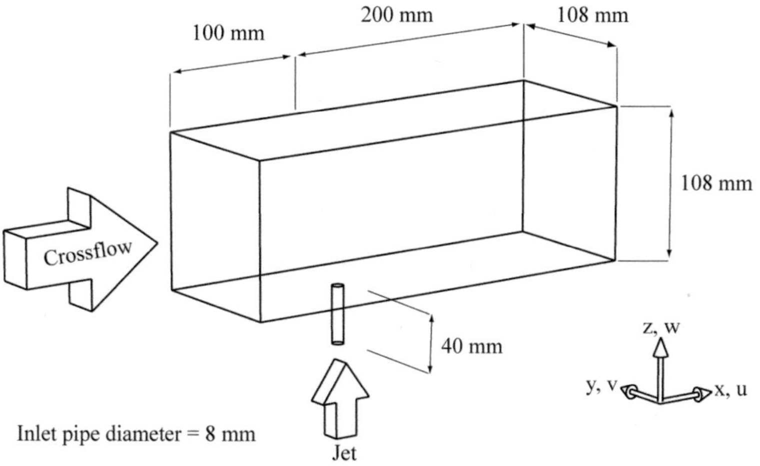

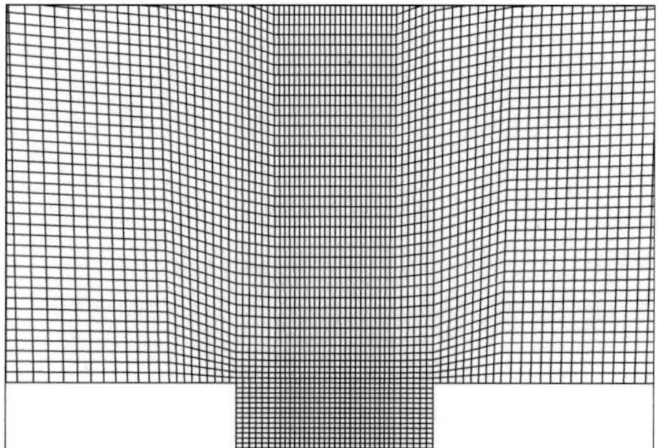

Figure 7.3: Overview of the computational domain with dimensions and the coordinate system employed, top. The grid at the symmetry plane near the jet inlet, bottom.

The typical y^+ value at the crossflow walls is 15 for the coarse and 12 for the fine grid, and at the pipe walls is 12 for the two grids.

For the unsteady simulations the time step was 2 microseconds long for the coarse grid and 1 microsecond long for the fine grid, yielding a maximum CFL number of approximately 0.3 for all simulations. An averaging time of at least 170 characteristic time units D/U_{cross} was employed for all simulations.

The grid dependence of the results was also explored. Figure 7.4 shows line plots of the mean velocity component U at the symmetry plane ($y/D = 0$) for LES simulations with the standard Smagorinsky model using the coarse and fine grids, compared to measurements for Case A. The effect of the grid resolution on the mean flow is very limited. Only the coarse grid shows some deviation at $x/D = 2$, which should be attributed to the coarser overall resolution than any near-wall flow effect.

The grid dependence of the results of the RANS simulations has been also investigated. The results using the coarse and fine grids were nearly identical, so the coarse grid was used for the RANS simulations.

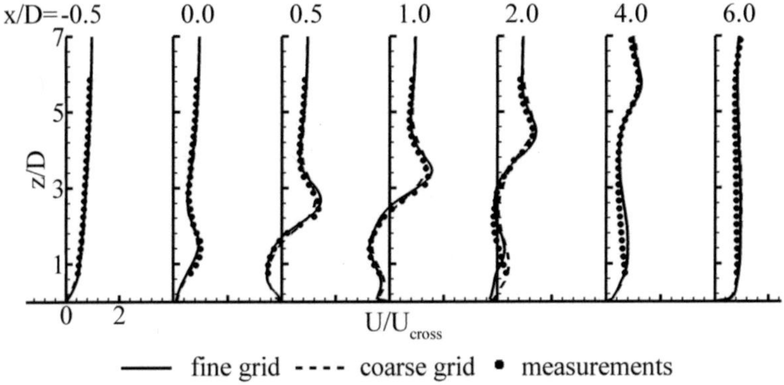

Figure 7.4: Line plots of the mean velocity component U/U_{cross} at the symmetry plane, $y/D = 0$, Case A.

Turbulence modeling

For the LES, two versions of the Smagorinsky subgrid scale turbulence model were available: the standard version with fixed model parameters and a dynamic version in which the model parameters are calculated from the smallest scales (see Section 5.4). To assess the performance of these two model versions, simulations with the boundary conditions of Case A and using the fine grid were performed. Line plots of the mean velocity component U/U_{cross} and normalized specific Reynolds stresses $\overline{u'u'}/U_{cross}^2$ at the symmetry plane ($y/D = 0$) for LES using the two models can be seen in Figure 7.5. The velocity profiles almost collapse, being difficult to distinguish between them. The results for $\overline{u'u'}$ are also very close, with the standard version predicting a slightly higher turbulence level. The very limited difference between the results of the two models is an indication that the role of the subgrid scale modeling is not decisive in this jet in crossflow configuration. The standard Smagorinsky model has the advantage of being less computationally intensive than the dynamic version, so this version was used in the simulations presented in the next sections.

Two models have been used for the RANS simulations, the standard k-ε and the SST models (see Section 5.2). For the unsteady RANS simulations, the SST model was preferred, as it produced the best results in the steady-state RANS simulations.

Boundary conditions

The description of turbulence at the inlet boundary conditions for LES is a known challenge [53, 51]. The first attempts using steady boundary conditions failed to predict the jet penetration correctly. To solve this problem, a turbulent boundary condition based on the work of Klein *et al.* [53], described in Section 5.7.3 was used. The mean velocity profiles from the RANS simulation have been applied to the inflow boundaries of the LES with superimposed turbulence fluctuations. The results show that this solution was very satisfactory, with good agreement between the LES and the measurements.

As already pointed out in the literature [79], the jet trajectory is influenced by both the jet velocity profile and the crossflow boundary layer. When the channel is small in comparison to the JIC, as in this work, the development of the crossflow boundary layer is even more important to the description of the boundary conditions since the velocity in the middle of the channel increases as the flow develops.

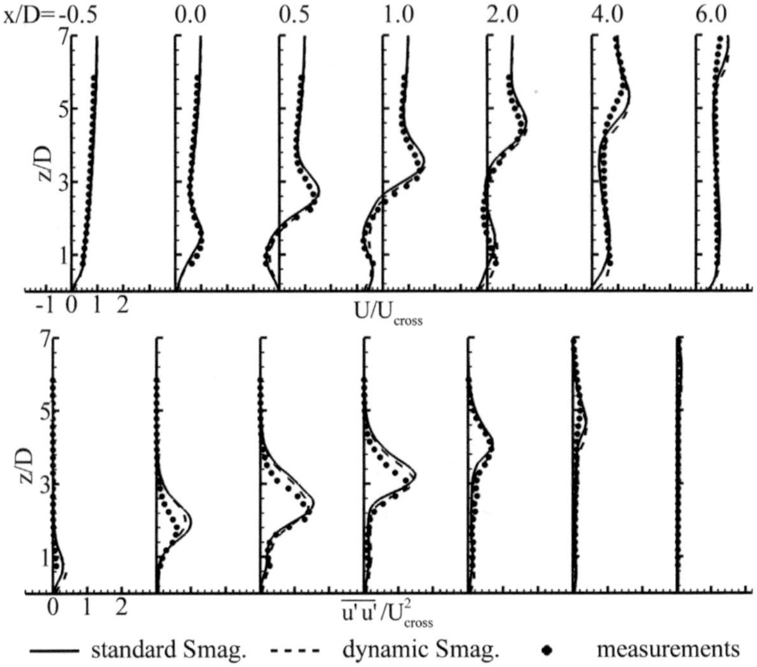

Figure 7.5: Line plots of mean velocity components U/U_{cross} and the specific Reynolds stress component $\overline{u'u'}/U_{cross}^2$ at the symmetry plane, $y/D = 0$, Case A.

To exemplify the sensitivity of the simulations to the crossflow and jet velocity boundary conditions, Figure 7.6 shows the comparison of the mean velocity U for two RANS simulations with the SST turbulence model: one with the correct velocity profiles, and one with the same bulk velocity but block profiles at the crossflow and jet inlets. The difference in the jet penetration is evident. Comparing the results with the measurements for Case A shows that the simulation with the correct velocity profiles has a better agreement than using the block profiles, as expected.

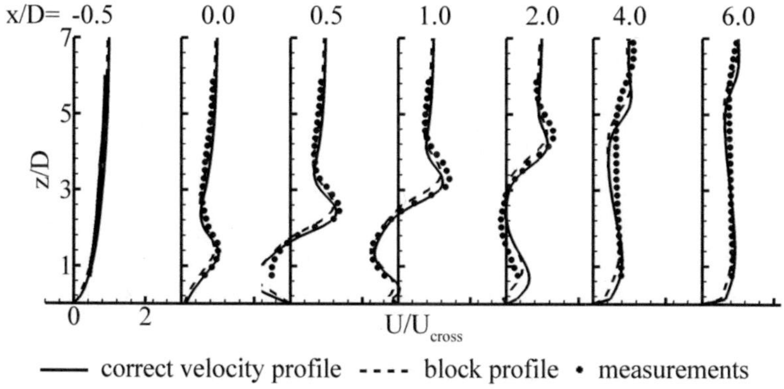

Figure 7.6: Line plots of the mean velocity component U/U_{cross} at the symmetry plane, $y/D = 0$, Case A.

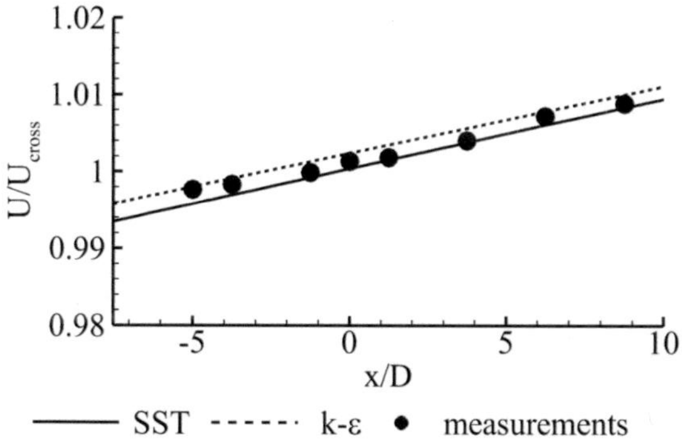

Figure 7.7: Developing velocity U/U_{cross} along the x direction, $z/D = 6.75$, $y/D = 0$, Case A.

In face of the sensitivity of the simulations to the boundary conditions, special attention was dedicated to them to allow the best level of agreement between the experimental data and the simulation. The crossflow boundary layer thickness could not be resolved using laser diagnostics because the reflections of the laser light caused by the walls interfered with the measurements. An alternative method was applied using the fact that the developing velocity profile of the crossflow depends on the growth of the boundary layer. Figure 7.7

shows measurements using laser Doppler anemometry (LDA) at the center of the channel along the x direction of the velocity component U of the crossflow alone without the jet compared to the RANS simulations. The curves are in good agreement with the measurements indicating that the growth of the boundary layer has been correctly simulated. Furthermore, the grid sensitivity of the developing velocity profile was checked. The grid resolution was enough to assure grid independent results.

Figure 7.8 shows the comparison between the measured and simulated velocity and turbulence profiles taken one diameter above the jet inlet ($z/D = 1$) and varying +/-1D along the x-direction. It can be seen that the measurements are very consistent; the peaks from the Reynolds stress components correspond to the regions of large mean velocity gradients for the two directions considered. The simulated velocity profiles in the x and z directions show good agreement with the measurements.

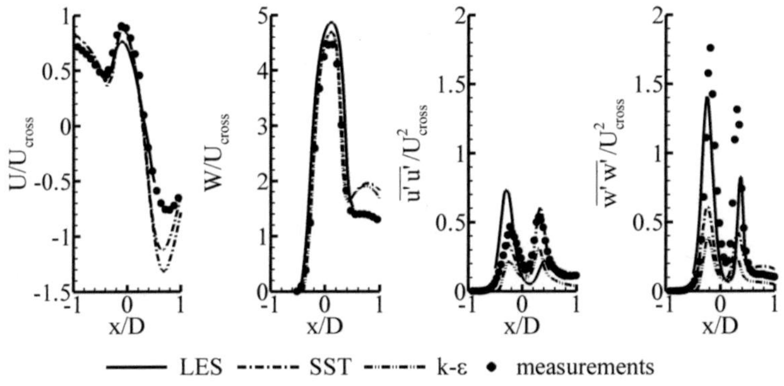

Figure 7.8: Profiles of velocity components, U/U_{cross} and W/U_{cross}, and specific Reynolds stresses, $\overline{u'u'}/U_{cross}^2$ and $\overline{w'w'}/U_{cross}^2$, one diameter above the jet inlet, $z/D = 1$, $y/D = 0$, Case A.

7.3 Results and discussion

7.3.1 Coherent structures

The ability of the different simulation strategies in predicting the formation and evolution of coherent structures is depicted in Figure 7.9, showing isosurface of 1% passive scalar for a snapshot of the LES and the unsteady SST (USST) simulation and for the steady-state result of the SST simulation, all for Case A.

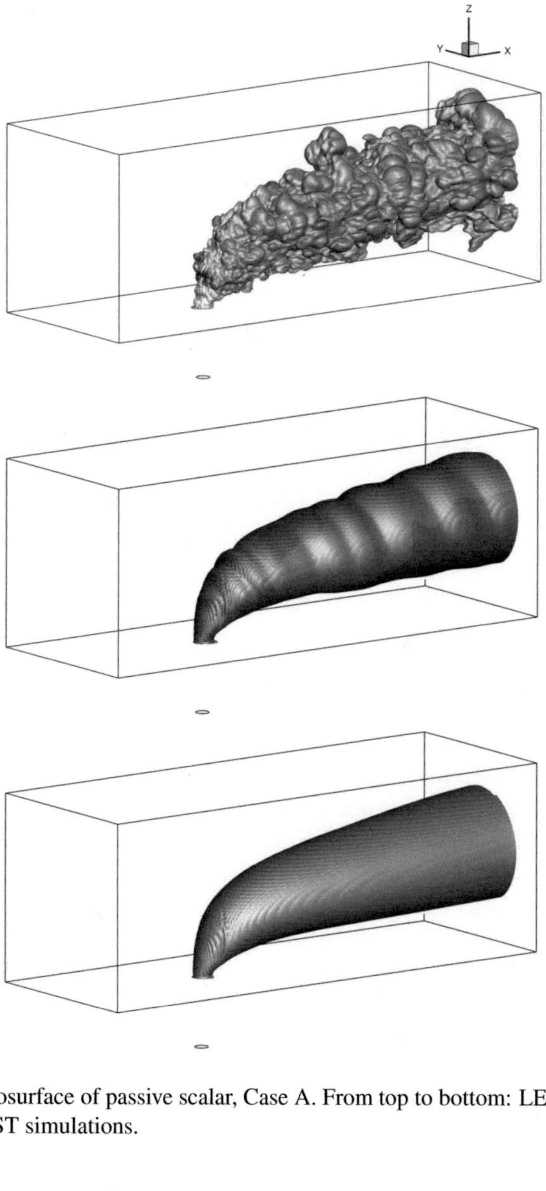

Figure 7.9: Isosurface of passive scalar, Case A. From top to bottom: LES, USST and SST simulations.

The isosurface of the LES is very contorted, indicating a high turbulence level. However, it is difficult to discern any specific coherent structure in this picture. Large-scale structures have also been predicted by the USST, resulting in the sinuous isosurface observed in Figure 7.9. It is evident, however, that the fluctuation level is much lower in comparison to the LES. The steady state SST simulation does not predict coherent structures at all, which results in a smooth passive scalar distribution.

It is difficult to discern any specific coherent structure from the LES in Figure 7.9. To this purpose, the isosurface of the pressure fluctuation $p - \overline{p}$ was used in the preceding chapter to highlight the formation of Kelvin-Helmholtz instabilities in a free jet (see Figure 6.17), and reveal the same structures in the jet in crossflow, as depicted in Figure 7.10. A constant value for the pressure fluctuation close to zero was used, which was chosen to enhance the presentation of the structures. Vortex rings can be observed just downstream of the jet inlet, which are due to the Kelvin-Helmholtz instability. The KH structures are not as dominant as in the case of the free jet, which can be attributed to two causes. Both jet and crossflow streams are turbulent, which makes the visualization of the structures more difficult. In addition, the bending of the jet in the crossflow direction acts as a destabilizing effect over the structures. Nevertheless, these structures are very similar to the structures found in the LIF cross section in Figure 3.10, although the huge difference in Reynolds number. The jet in Figure 3.10 is laminar, with a Reynolds number of 1 600, while the jet Reynolds number of the LES is 19 200.

Further structures can be analyzed using streamtraces. Figures 7.11 and 7.12 show two views of the same streamtraces originated at the jet inlet, following the instantaneous velocity field. The Figure 7.12 shows a view of the channel from the outlet towards the inlet. It is especially interesting to see how the counter-rotating vortex pair is clearly visible. Figure 7.13 show streamtraces at the symmetry plane near the jet inlet. Directly upstream of the jet, the location where the streamtraces flow together is the horseshoe vortex. This structure is remarkably similar to the horseshoe vortex formed at a cylinder-wall junction.

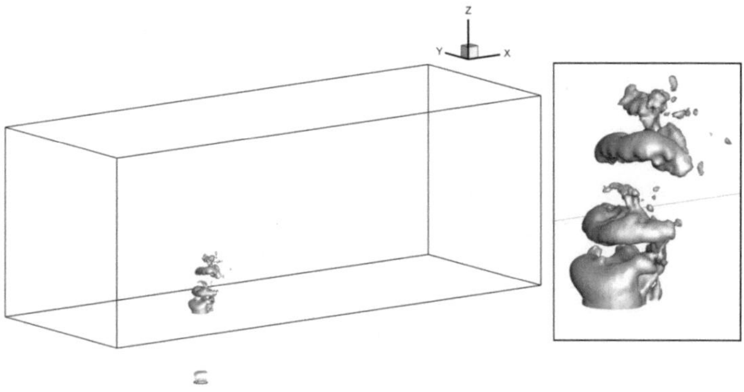

Figure 7.10: Isosurface of pressure fluctuation, LES of Case A.

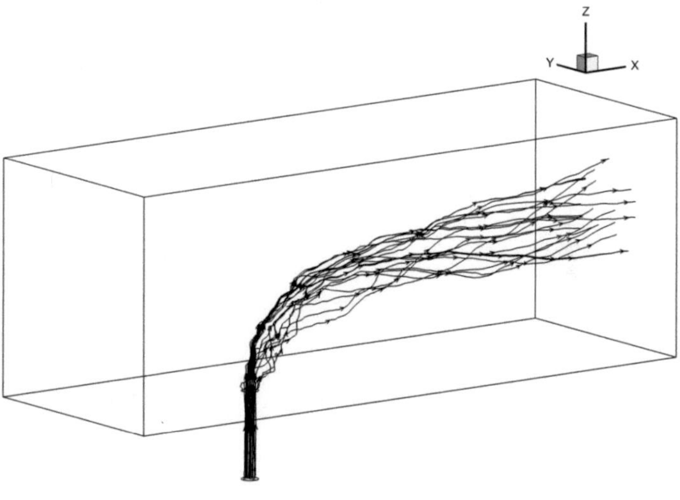

Figure 7.11: Streamtraces originated at the jet inlet, LES of Case A.

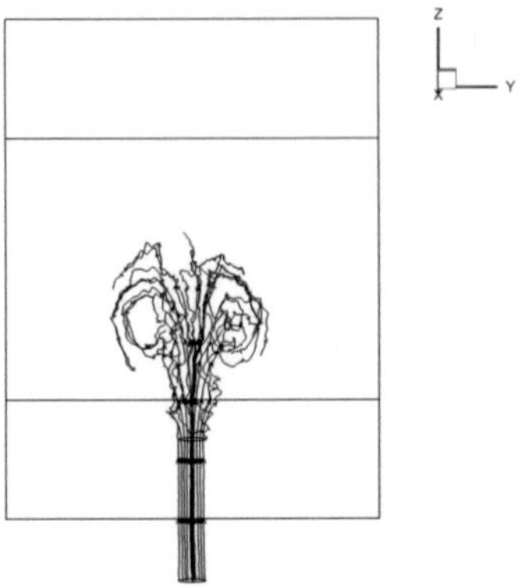

Figure 7.12: Streamtraces originated at the jet inlet, view from the outlet towards the inlet, LES of Case A.

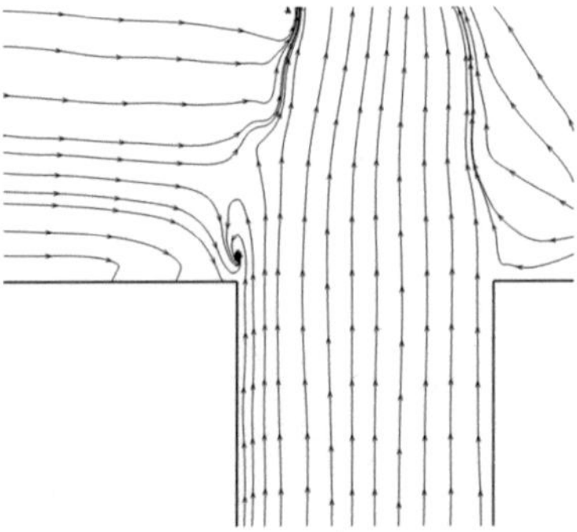

Figure 7.13: Streamtraces at the symmetry plane, $y/D = 0$, LES of Case A.

As in the case of the free jet, analyzing the diffusion across the superlayer can help to understand the discrepancy between the velocity and passive scalar fields. In order to calculate the diffusion from the LES, the superlayer was approximated by the surface determined by the isosurface of 1% passive scalar, which is the same used to define the intermittency function (see Figure 7.9, top). To allow a direct comparison, the velocity component w non-dimensioned by the bulk jet velocity U_{jet} is employed along the passive scalar c, as both are transported into the crossflow by the jet and span between zero and one.

The diffusion flux across the superlayer of velocity J_w and of the passive scalar J_c have been defined by Equations (6.7) and (6.8), reproduced here for convenience

$$J_w = \nu \frac{\partial (w/U_{jet})}{\partial x_i} \cdot \vec{n} \tag{7.1}$$

$$J_c = D \frac{\partial c}{\partial x_i} \cdot \vec{n} \tag{7.2}$$

Integrating over the whole isosurface of a snapshot of the LES of Case A resulted in a dimensionless velocity flow rate of $2.64 \cdot 10^{-6}$ m^3/s and a passive scalar flow rate of $26.59 \cdot 10^{-6}$ m^3/s, indicating that the passive scalar gradient is greater than the velocity gradient across the superlayer. As in the case of the free jet, the greater passive scalar flow rate across the superlayer corroborates with the fact that the spreading rate of the passive scalar is larger than the spreading rate of the velocity.

7.3.2 Wake shedding frequency

The flow at the lee side of the jet in crossflow (lee side is the side that is sheltered from the crossflow) produces alternate vortices shed in a similar way to the Kármán vortex street formed behind solid bluff bodies, with has been observed by many authors [28, 50, 77, 125]. The fluid at the boundary layer is channeled out from the boundary through the so-called *wake vortices*. These vortices do not form a real wake, however, as there is no detachment of the flow [77].

Fric and Roshko [28] have characterized the jet wake using the wake Strouhal number, St, which is a dimensionless number describing the oscillation frequency of the flow. It is known that the Kármán vortex street has a characteristic Strouhal number of about 0.2 in the same range of Reynolds numbers as they used in this work. The Strouhal number as defined by Fric and Roshko reads

$$\mathrm{St} = \frac{fD}{U_{cross}}, \tag{7.3}$$

where f is the frequency of vortex shedding, D is the jet diameter and U_{cross} is the bulk crossflow velocity. Figure 7.14 shows values of St measured by Fric and Roshko. The repeatability of the measurements (Figure 7.14.a) is good at a velocity ratio of 4; however, it shows a broad spectrum at a velocity ratio of 3.

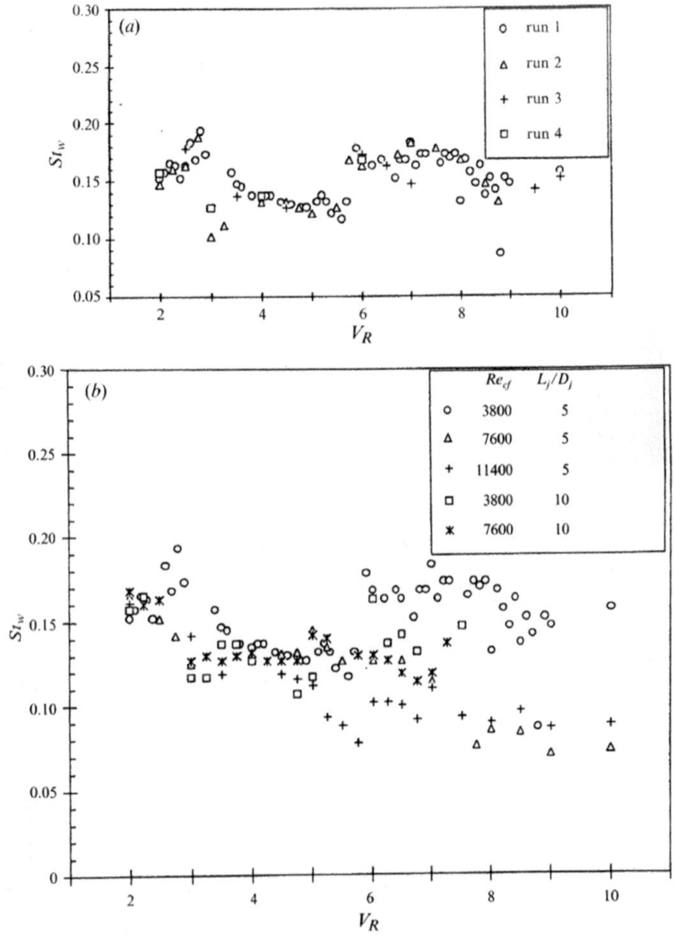

Figure 7.14: Measurements of Fric and Roshko [28] of the wake Strouhal number St (St_w). Degree of repeatability of the measurements (a); data for several combinations of Re_{cf} and velocity ratios V_R, measured at $x/D = 3.5$, $y/D = 1.5$ and $z/D = 0.5$. Reproduced from [28].

For a range of Reynolds numbers defined as $\mathrm{Re}_{cf} = U_{cross}D/\nu$ from 3 800 to 11 400 and velocity ratios between 3 and 4, the Strouhal number measured by Fric and Roshko [28] lies between 0.12 and 0.16, being systematically higher for the experiments with lower Reynolds numbers.

An analysis of the wake frequency has been done for the LES from Case A. The Reynolds number defined as above is 4 650, and the velocity ratio is 4.15, which is a good match to the results of Fric and Roshko [28]. Figure 7.15 shows the collection of monitor points that have been placed in the computational domain; they are located on only one half of the computational domain, taking advantage of the symmetry of the geometry. The time evolution of the passive scalar of each monitor point was analyzed using fast Fourier transforms; the monitor point with the cleanest signal was located at $x/D = 8$, $y/D = 3$ and $z/D = 1$ and is also indicated in Figure 7.15. Figure 7.16 shows the time evolution of the passive scalar for the selected monitor point. The result of the fast Fourier transform is depicted in Figure 7.17. The analysis found a vortex shedding frequency of 224 Hz, leading to a Strouhal number of 0.197. The value is slightly higher than the experimental results of Fric and Roshko; nevertheless the agreement is good considering the differences in the experimental setup and boundary conditions. This is another good indication that the large-scale structures of the flow have been resolved correctly by the LES.

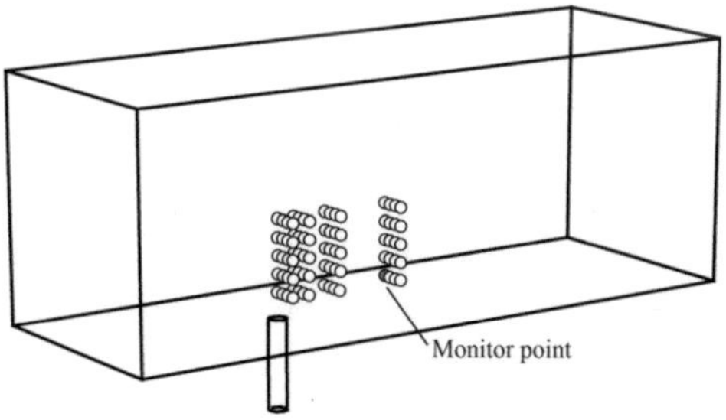

Figure 7.15: Overview of the computational domain with the monitor points. The monitor point with the cleanest signal is indicated with gray color.

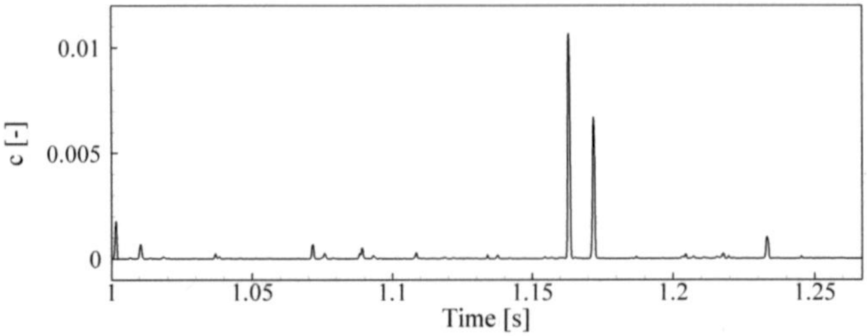

Figure 7.16: Time evolution of the passive scalar of the monitor point located at $x/D = 8$, $y/D = 3$ and $z/D = 1$.

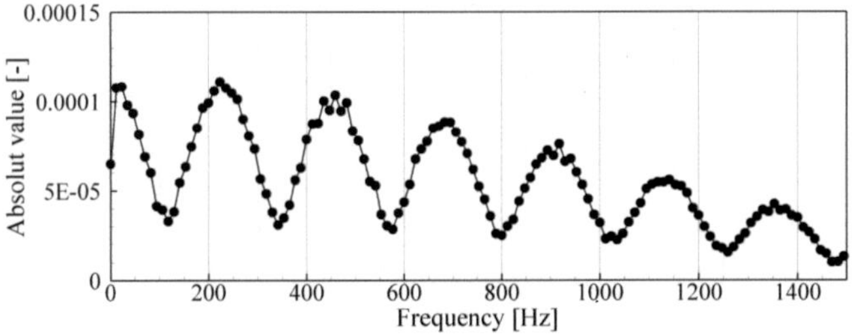

Figure 7.17: Fast Fourier transform of the passive scalar signal of the monitor point located at $x/D = 8$, $y/D = 3$ and $z/D = 1$.

7.3.3 Probability Density Function at intermittent regions

Pope [93] examined one-point Probability Density Functions (PDFs) measured in shear flows, with the conclusion that, in homogeneous shear flow with a uniform mean scalar gradient, the joint PDF of velocity and the scalar is joint normal; whereas in free shear flows the PDFs are not Gaussian. Figure 7.18, reproduced from [93], shows the standardized marginal PDFs of the velocity components u, v, w and the passive scalar ϕ in a homogeneous shear flow. The profiles follow closely the Gaussian profile. Pope also shows that velocity-velocity and velocity-scalar joint PDFs of the same flow are accurately described by joint-normal distributions. Analyzing free shear flows Pope found a quite different behavior. Figure 7.19, reproduced from [93], shows the scalar PDFs $f_\phi(\psi, \xi)$ in a mixing layer, where ψ and ξ are the sample space variables of ϕ and t, respectively. The scalar values in the two streams are $\phi = 0$ and $\phi = 1$. As the passive scalar is a bounded variable (see Equation (2.4)), $\phi(x_i, t)$ must lie everywhere between zero and unity; consequently the probability f_ϕ if zero for any value that is not in this interval.

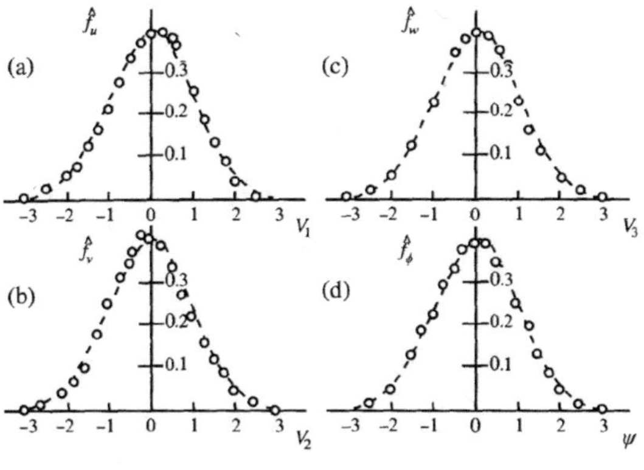

Figure 7.18: Standardized PDFs of (a) u, (b) v, (c) w and (d) ϕ in homogeneous shear flow. Dashed lines are standardized Gaussians. Reproduced from [93], values from measurements by [109].

As may be seen in Figure 7.19, in the center of the layer the PDF is broad and roughly bell-shaped, spanning the entire range of values. As the measurement location moves toward the high-speed stream, the PDF moves to higher values of ψ and develops a spike of increasing magnitude at the upper bound $\psi = 1$, assuming a clearly non-Gaussian shape.

The following analysis shows that the LES is able to reproduce the non-Gaussian behavior in the JIC. The simulation of Case A has been analyzed.

Figure 7.20 shows PDFs of passive scalar c and the velocity components u, v and w, calculated at Point A, which is situated in the recirculation region on the lee side of the jet ($x/D = 1$, $y/D = 0$ and $z/D = 2$, see Figure 7.21), where the influence of the intermittency is expected to be small. A total of 130 000 time steps were processed and analyzed. Apart from small deviations, the PDFs in Figure 7.20 are roughly bell-shaped.

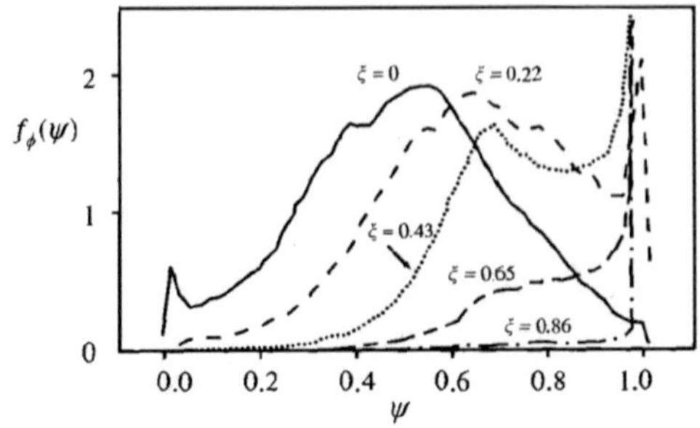

Figure 7.19: PDFs of a conserved passive scalar in the self-similar temporal mixing layer at various lateral positions. Reproduced from [93], values from direct numerical simulations by [102].

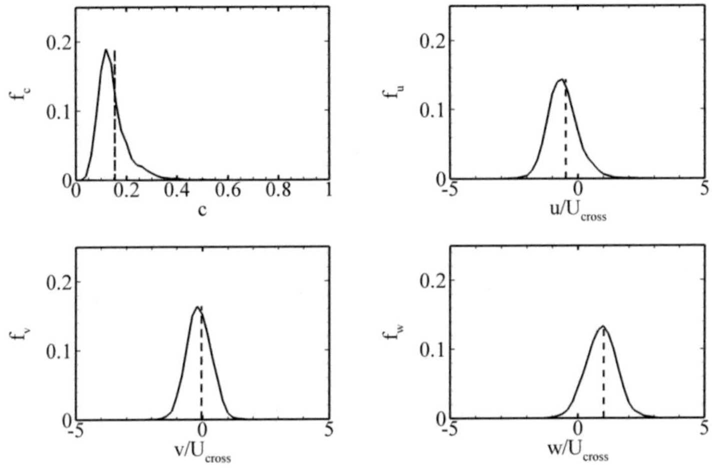

Figure 7.20: PDFs of passive scalar c and the velocity components u/U_{cross}, v/U_{cross} and w/U_{cross} at Point A. The dashed lines represent the mean values.

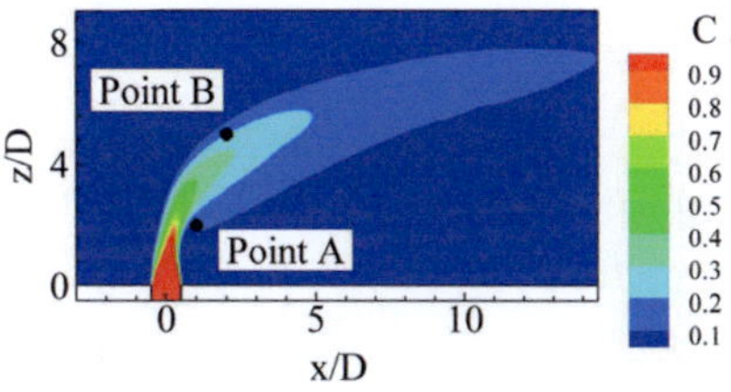

Figure 7.21: Two dimensional map of mean passive scalar C along with the position of Points A and B used in PDF analysis.

Analyzing now the results for the Point B placed at the intermittent region of the jet leads to different results. Point B is placed at $x/D = 2$, $y/D = 0$ and $z/D = 5$, close to the region with strong formation of coherent structures (see Figure 7.21). The statistics of u and v in Figure 7.22 remain roughly bell-shaped, while the distributions of c and w depart clearly from the bell shape, with a much broader distribution. The PDF of c exhibits a double-peak distribution, which agrees with similar PDFs of jets shown, e.g., by Dahm and Dimotakis [19].

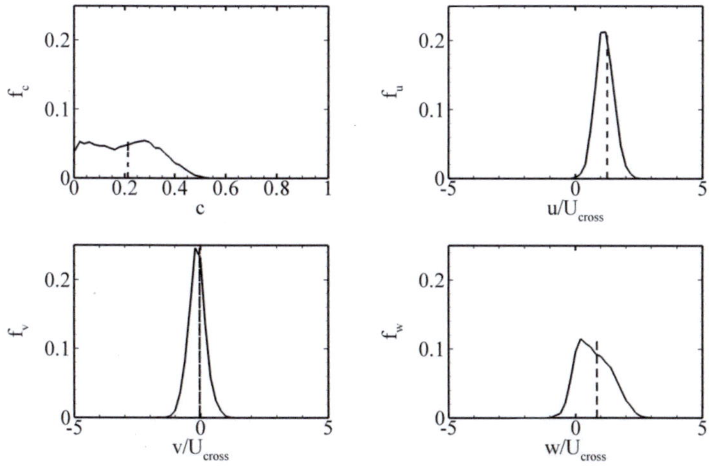

Figure 7.22: PDFs of passive scalar c and the velocity components u/U_{cross}, v/U_{cross} and w/U_{cross} at Point B. The dashed lines represent the mean values.

For the Point A, Figure 7.23 shows a three-dimensional representation of the joint PDF of the velocity components *u* and *v*. The joint PDF is bell-shaped, as the marginal PDF of *u* and *v* in Figure 7.20. A two-dimensional representation of the same joint PDF can be seen in Figure 7.24, along with joint PDFs of other variable pairs. There is no strong departure of the bell shape for any variable pair.

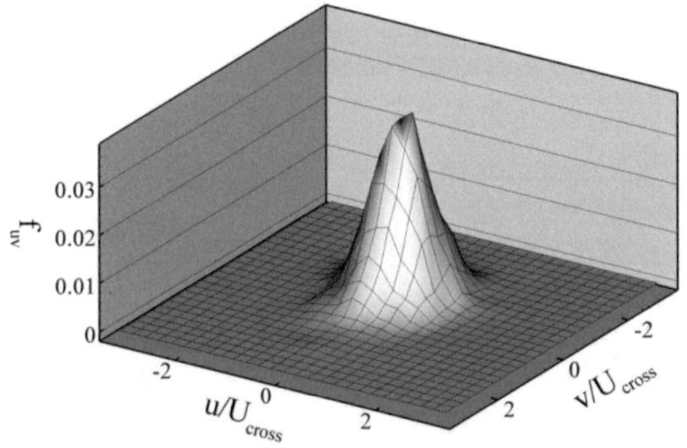

Figure 7.23: Three-dimensional representation of the joint PDF of the velocity components *u* and *v* at Point A.

The joint PDFs at Point B are depicted in Figure 7.25. While the joint PDFs of the velocity pair are bell-shaped, the velocity-scalar joint PDFs show a significant departure of the bell shape. The *u*/*c* and *v*/*c* pairs show a double-peak distribution which agrees well with the marginal PDF of *c* in Figure 7.22. The joint PDF of *w* and *c* has a double peak distribution, with an elongated shape with an angle of approximately 45 degrees. It means that *w* and *c* are strongly correlated at this point; the joint PDF of perfectly correlated variables would be a straight line with an angle of 45 degrees, in a two-dimensional representation with the same scale for both variables. The strong correlation between *w* and *c* can be explained by the fact that both passive scalar *c* and the *w* velocity component are characteristics of the jet, which are transported to this location mainly by the large-scale coherent structures. Quantities transported by turbulence typically do not exhibit such strong correlation, as seen in Figure 7.20, where the influence of coherent structures is limited and the turbulent transport prevails. It is interesting to note here that neither the double-peak distributions nor the strong correlation between *w* and *c* can be simulated by RANS turbulence models, which implicitly assume joint-normal distributions for velocity-velocity and velocity-scalar joint PDFs.

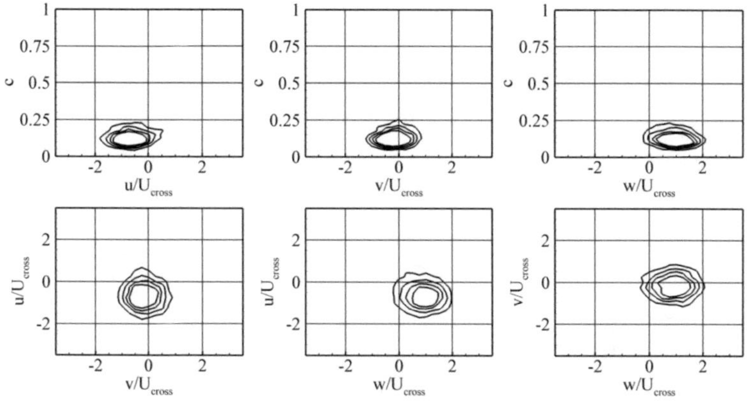

Figure 7.24: Contour plots of joint PDFs of *u* and *c*, *v* and *c*, *w* and *c*, *u* and *v*, *u* and *w*, *v* and *w* at Point A. Contour values are 0.02, 0.015, 0.01 and 0.005.

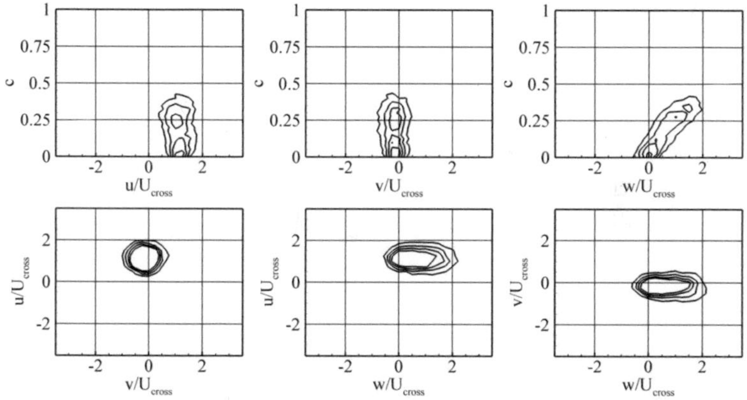

Figure 7.25: Contour plots of joint PDFs of *u* and *c*, *v* and *c*, *w* and *c*, *u* and *v*, *u* and *w*, *v* and *w* at Point B. Contour values are 0.02, 0.015, 0.01 and 0.005.

7.3.4 Flow, turbulence and mixing

In this section the results of the simulations will be compared to the measurements of Case A, described in reference [31]. Figure 7.26 shows the position of the symmetry plane ($y/D = 0$) and the horizontal plane ($z/D = 1.5$) used in the measurements. The results of the following simulations are presented: a large eddy simulation (LES) using the fine grid and the standard Smagorinsky subgrid-scale model; an unsteady RANS simulation using the SST turbulence model (USST) and steady state RANS simulations using the SST and k-ε turbulence models. The coarse grid was used in all RANS simulations.

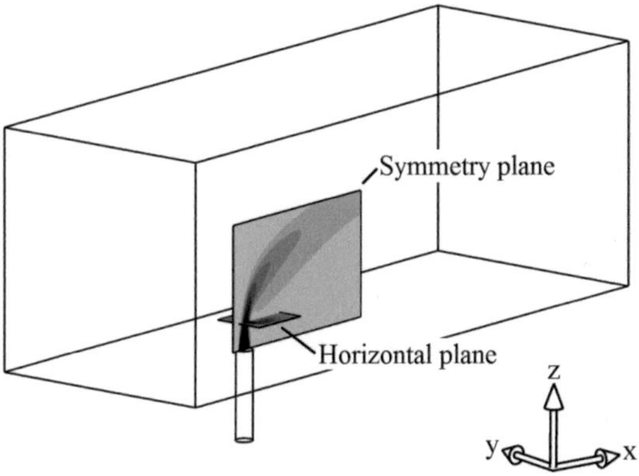

Figure 7.26: Overview of jet in crossflow setup showing the symmetry plane ($y/D = 0$) and the horizontal plane ($z/D = 1.5$) used in the measurements.

It will be seen that the results of the LES have the best agreement with the measurements. However, the good agreement comes with a high computational cost. This cost can be reduced applying unsteady RANS (URANS) simulations instead of LES. In the URANS approach, the whole spectrum of turbulence is modeled, and only the unsteadiness that comes from the coherent structures is resolved directly. It will be shown that the steady state simulations with the SST turbulence model are superior to that with k-ε. The SST turbulence model was consequently used in the unsteady simulations. The use of steady state simulations using RANS turbulence models instead of the unsteady LES or URANS simulations reduce the computational cost even more. In this approach, the whole spectrum of turbulence is modeled, and the unsteadiness originated from the coherent structures is ignored. The performance of two turbulence models, k-ε and SST, was assessed. Annex A.3 shows a detailed comparison of the computational cost for the different simulations.

Prior to the discussion about the quality of the simulation of mixing, the velocity field and the overall flow features must be considered, as they constitute a prerequisite for the correct description of the mixing phenomena.

The profiles in Figure 7.8 are a good indication of the quality of the boundary conditions used in the simulations. Both mean velocity components U and W predicted by the LES agree well with the measurements. The agreement of the steady state RANS is also good in the jet core, however is degraded starting at $x/D = 0.5$, the location where the pipe ends. Comparing the specific Reynolds stress components $\overline{u'u'}$ and $\overline{w'w'}$ from the steady state RANS simulations, while presenting good qualitative agreement, they show an overall underestimated turbulence level. The LES predictions are in better to the experimental data. Between the RANS simulations, the SST model yields a better match to the data than the k-ε model.

The two dimensional plots in Figure 7.27 are a good starting point for the discussion of the results. In this Figure, PIV measurements and the results of the simulations for the mean velocity component U and the specific Reynolds stress component $\overline{u'u'}$ at the symmetry plane ($y/D = 0$) are shown. The jet is mounted in the z direction, so the jet fluid that enters in the channel has a significant W velocity component and a nearly zero U velocity component. As the jet flows into the crossflow, it is bent in the crossflow direction and U increases, creating the region of high U situated at $x/D = 1$ and $z/D = 4$. With the development of the jet, crossflow fluid is entrained, and U decreases continuously. In the lee side of the jet, a recirculation zone takes place. This region has negative U values, which is represented by the blue color in the contours. The recirculation region is an essential part of the structure of the jet in crossflow, however it is difficult to measure and simulate adequately. It is interesting to note that the recirculation region and the region with maximum U are very close to each other. The velocity gradient in this region is therefore very high, and very high production of turbulence is expected. The contour of $\overline{u'u'}$ confirms the expectation, with the locus of maximum $\overline{u'u'}$ situated in the region of high velocity gradient.

Comparing the results of the simulations with the measurements in Figure 7.27, the LES has clearly the better agreement. Both velocity and turbulence fields are very well reproduced, having the field maxima at the same position and having almost the same intensity as the measurements. Although the position of the jet is also good reproduced by the other simulations, the maximum values of U are slightly underpredicted. The predictions of $\overline{u'u'}$,

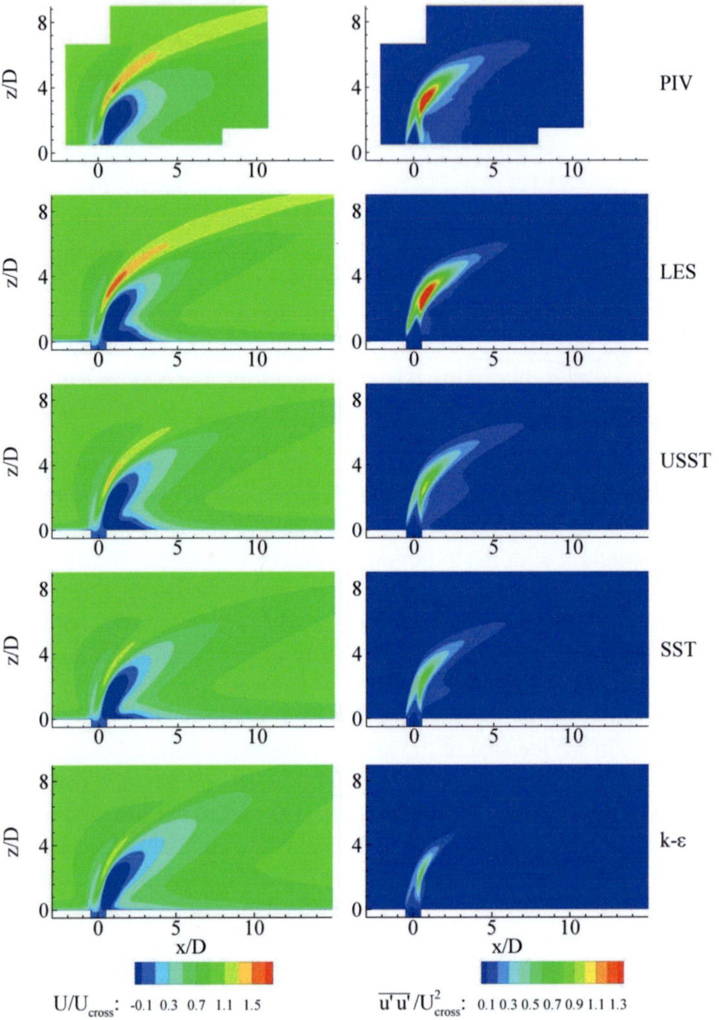

Figure 7.27: Two dimensional maps of mean velocity components U/U_{cross} specific Reynolds stress component $\overline{u'u'}/U_{cross}^2$ at the symmetry plane, $y/D = 0$, Case A. From top to bottom: PIV measurements, LES, unsteady SST, steady state SST and steady state k-ε simulations.

on the other hand, differ substantially from the measurements. The Reynolds stress components of the steady state RANS simulations have been calculated using the Boussinesq hypothesis, Equation (5.4). For the unsteady SST (USST) simulation, the Reynolds stress components have been calculated summing up the contributions of the turbulence modeling, using the Boussinesq hypothesis, and the root mean square of the velocity fluctuation about the mean calculated directly during the simulation:

$$\underbrace{-\overline{u_i'u_j'}}_{\text{total, URANS}} = \underbrace{2\nu_t S_{ij} - \frac{2}{3}k\delta_{ij}}_{\text{Boussinesq}} - \underbrace{\overline{u_i'u_j'}}_{\text{resolved}} \; . \tag{7.4}$$

While the measurements show a maximum value of $\overline{u'u'}/U^2_{cross}$ at the symmetry plane of 1.89, the prediction of the LES is 1.60, of the USST is 1.07, of the SST simulation is 0.90 and of the k-ε simulation is only 0.68. This fact can be explained by the presence of large-scale coherent structures in the jet in crossflow, which induce a statistically non-stationary behavior (see Section 2.4 for more details). The measurements were evaluated using simple time averages, so the Reynolds stress tensor represents the whole velocity fluctuation about the time averaged mean velocity field. This implies that these fluctuations contain the contribution of two different phenomena: the turbulent fluctuations and the unsteadiness of the mean flow about the time-average created by the coherent structures. It has been shown in the preceding section that the LES was able to resolve the major coherent structures of the jet in crossflow; it is confirmed by the very good agreement of the fluctuation level of the LES with the measurements. It is argued that the reason for the low $\overline{u'u'}$ levels predicted by the USST simulation is the insufficient description of the coherent structures. This simulation could resolve a fraction of the unsteady character of the flow; however, a significant fraction remained unresolved. This unresolved fraction of the fluctuation level contributes to the low levels predicted by the USST simulation in comparison with the measurements. As the coherent structures are not resolved at all by steady state RANS simulations, their fluctuation level contains only the contribution of the turbulence and is consequently lower than both unsteady simulations and measurements.

Line plots allow the quantitative comparison between measurements and simulations. Figures 7.28 and 7.29 show the comparison of the mean velocity components U and W, and the specific Reynolds stress components $\overline{u'u'}$ and $\overline{w'w'}$ between values extracted from the PIV measurements at the symmetry plane ($y/D = 0$) and the simulations. One of the most critical parameters of the jet in crossflow for engineering applications is the jet penetration, which is defined in this work as the locus of maximum mean velocity component U.

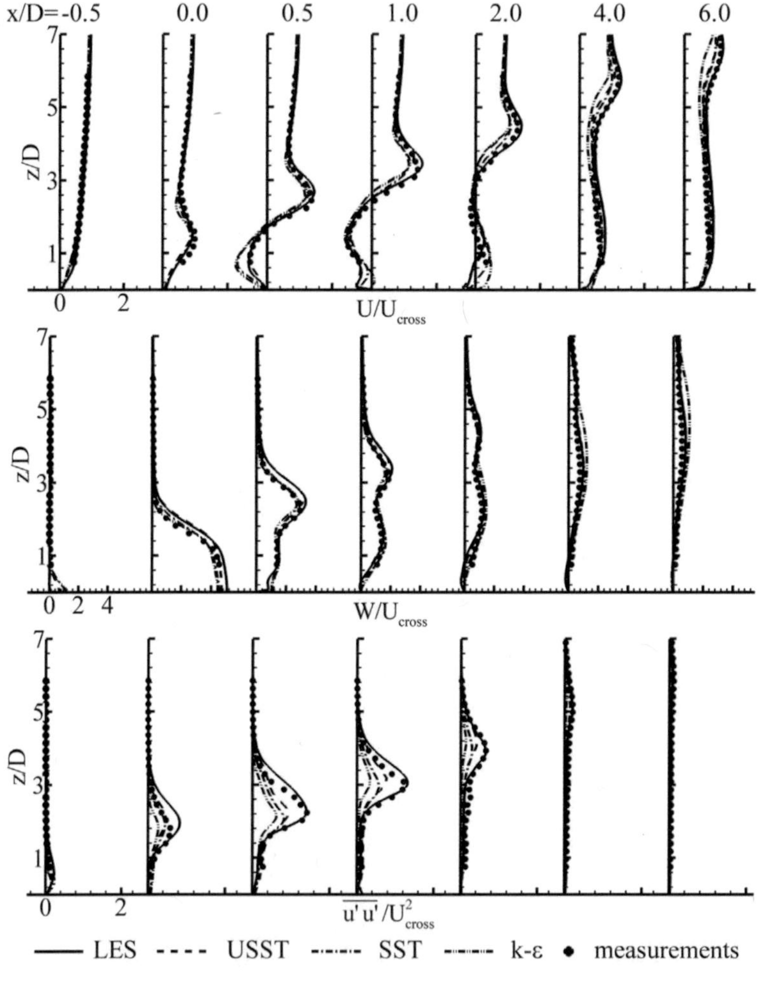

Figure 7.28: Line plots of mean velocity components U/U_{cross} and W/U_{cross}, and the specific Reynolds stress component $\overline{u'u'}/U^2_{cross}$ at the symmetry plane, $y/D = 0$, Case A.

The profiles of U nearly collapse, indicating that the jet penetration is well represented by all simulations except the RANS simulation with the k-ε turbulence model, which results deviate from the measurements from $x/D = 4.0$ and further downstream. Comparing the profiles for U and W in more detail, the LES shows very good agreement in all positions, even near the bottom wall, a region known to be difficult to simulate accurately due to the recirculating flow present there. The three RANS simulations show similar results, with velocity magnitudes that are slightly lower than the measurements in the region where the jet is located. The agreement of the LES results with the measurements for the specific Reynolds stress components $\overline{u'u'}$ and $\overline{w'w'}$ is very good in both magnitude and location, with only a slightly overpredicted $\overline{u'u'}$ level at $x/D = 0$. As seen in Figure 7.27, the USST simulation shows values that are consistently lower than both measurements and LES for $\overline{u'u'}$ and $\overline{w'w'}$, and the results of steady state RANS simulations are even lower. Following the results obtained with the free jet, the turbulence level predicted by the k-ε model is consistently lower than the one predicted by the SST model.

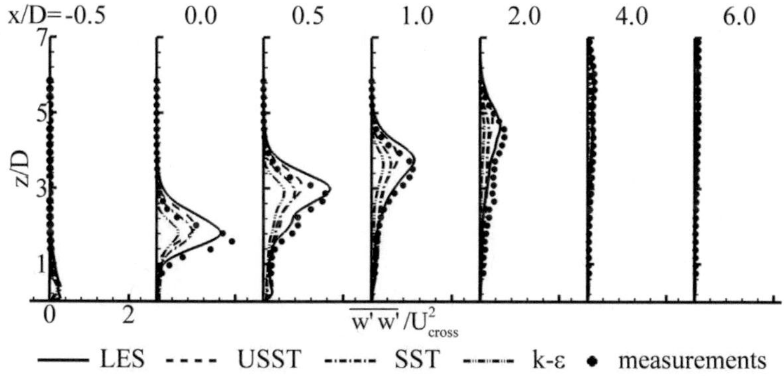

Figure 7.29: Line plots of the specific Reynolds stress component $\overline{w'w'}/U_{cross}^2$ at the symmetry plane, $y/D = 0$, Case A.

The results of the horizontal planes corroborate the results obtained for the symmetry plane. Figure 7.30 shows two dimensional plots of the PIV measurements at a plane located 1.5 D above the jet inlet ($z/D = 1.5$) compared with the simulations for the mean velocity component U and the specific Reynolds stress components $\overline{u'u'}$ and $\overline{v'v'}$. All simulations agree well with U,

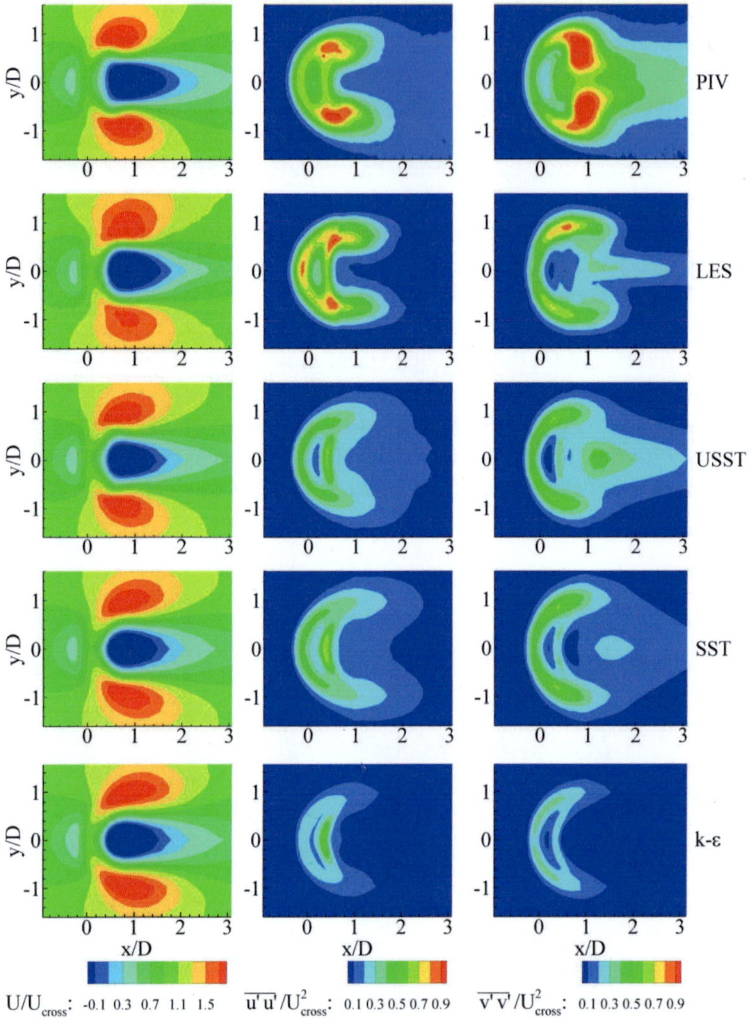

Figure 7.30: Two dimensional maps of mean velocity components U/U_{cross} and V/U_{cross} and specific Reynolds stress components $\overline{u'u'}/U_{cross}^2$, $\overline{v'v'}/U_{cross}^2$ and $\overline{u'v'}/U_{cross}^2$, $z/D = 1.5$, Case A. From top to bottom: PIV measurements, LES, unsteady SST, steady state SST and steady state k-ε simulations.

however with a slightly underpredicted length of the recirculation zone. The experiments show $\overline{u'u'}$ and $\overline{v'v'}$ with a distinct character as two peak values can be observed on the sides of the jet; in the case of $\overline{u'u'}$ directly on the jet side, and in the case of $\overline{v'v'}$ just downstream of the jet. The LES predicts $\overline{u'u'}$ in very good agreement with the measurements; both the pattern and the levels agree well. The agreement is also good with $\overline{v'v'}$; the location of the peaks is slightly moved in the upstream direction and the peak values are not so pronounced as in the experiments. The measurements show maximum values of $\overline{u'u'}/U^2_{cross}$ and $\overline{v'v'}/U^2_{cross}$ of 1.03 and 1.15, respectively, while the LES predicted 0.92 and 0.85. The agreement of the unsteady SST simulation is not so good, as the two peak values on the sides of the jet for $\overline{u'u'}$ were not predicted by the simulations. The same applies to $\overline{v'v'}$, with inferior quantitative and quantitative agreement. The overall fluctuation level is lower than the measurements; the unsteady SST simulation predicted peak values of 0.62 for $\overline{u'u'}/U^2_{cross}$ and 0.61 for $\overline{v'v'}/U^2_{cross}$. The steady state RANS simulations predicted peak values of $\overline{u'u'}$ directly upstream and downstream the jet, while $\overline{v'v'}$ shows only one peak region at the upstream portion of the jet, which do not agree with the measurements. In addition, the predicted values are significantly lower than the measurements. For example, the peak values predicted by the SST simulation and the k-ε simulation are 0.75 and 0.39, respectively.

A more detailed comparison of the mean velocity component U and the specific Reynolds stress component $\overline{u'u'}$ at the symmetry plane ($y/D = 0$) can be seen in Figure 7.31 for three different z/D positions (1.5, 3 and 4.5). The results confirm the very good agreement of the LES with the measurements for both U and $\overline{u'u'}$. The mean velocity U of the unsteady SST simulation shows good agreement with the measurements at $x/D = 1.5$ and 3.0, however the position of the jet at $x/D = 4.5$ is slightly shifted downstream. The fluctuation level shows good qualitative agreement, however the predicted levels are clearly lower than the experimental results, especially at $x/D = 3.0$. The steady state SST simulation shows a good agreement with the measurements for U, while the simulation using the k-ε model predicted the jet to be narrower than in the measurements. Following the results seen before, the SST and k-ε simulations show consistently lower $\overline{u'u'}$ levels than in the measurements. Only at $z/D = 1.5$, which is close to the jet inlet into the crossflow, the $\overline{u'u'}$ levels agree better with the measurements. In this region the coherent structures have not evolved sufficiently and could affect only marginally the results, which can explain the good agreement. In regions where the coherent structures have already developed, the agreement is significantly degraded.

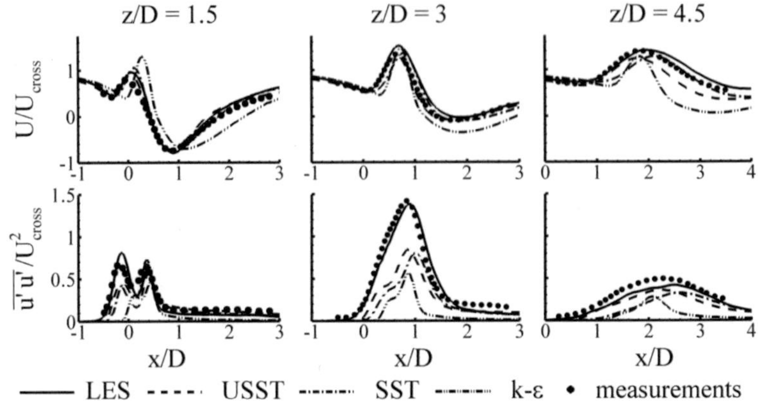

Figure 7.31: Line plots of velocity component U/U_{cross} and specific Reynolds stress component $\overline{u'u'}/U_{cross}^2$ at the symmetry plane, $y/D = 0$, Case A.

Figure 7.32 shows line plots of the mean passive scalar C and the intermittency function γ at the symmetry plane for the simulations. The intermittency function γ has been defined in Section 4.1 using the passive scalar; the classical definition using the turbulence is not appropriate for the JIC, as both flows are turbulent, making the definition with turbulence ambiguous. Unfortunately there are no measurements of C at the symmetry plane for Case A; the comparison of the simulations with experimental data at the horizontal plane will be shown in the following Figures 7.33 and 7.34. The C profiles show more pronounced differences between the different simulations than the mean velocity components U and W in the same locations in Figure 7.28. The LES profiles of C show a broader distribution than the profiles of the USST, SST and k-ε simulations, in this order, which can be more clearly seen in the profiles at $x/D = 0.5$, 1.0 and 2.0. This behavior agrees well with what can be learned from the profiles of intermittency factor γ. The γ function represents the fraction of time in which each point remains in the jet fluid, and its distribution is closely related to the amount of large-scale coherent structures that have been resolved by the simulation. Steady state RANS simulations do not resolve coherent structures at all, and the γ profiles assume the shape of a step function, because in this case the intermittency factor γ is equal to the intermittency function I, assuming the value zero when C is smaller than 0.01 and the value unity otherwise (for more detail see Section 4.1). In all x/D

positions, the regions where $\gamma = 1$ for the k-ε simulation are narrower than the SST simulation, which is a consequence of the narrower C profiles predicted using the k-ε model. Moving to the unsteady simulations, the γ distribution of the LES has less steep gradients than of the USST simulation, especially at the jet borders. It indicates that the LES is able to resolve a greater fraction of the large-scale coherent structures than the USST, as the steep gradients are typical for the steady state simulations.

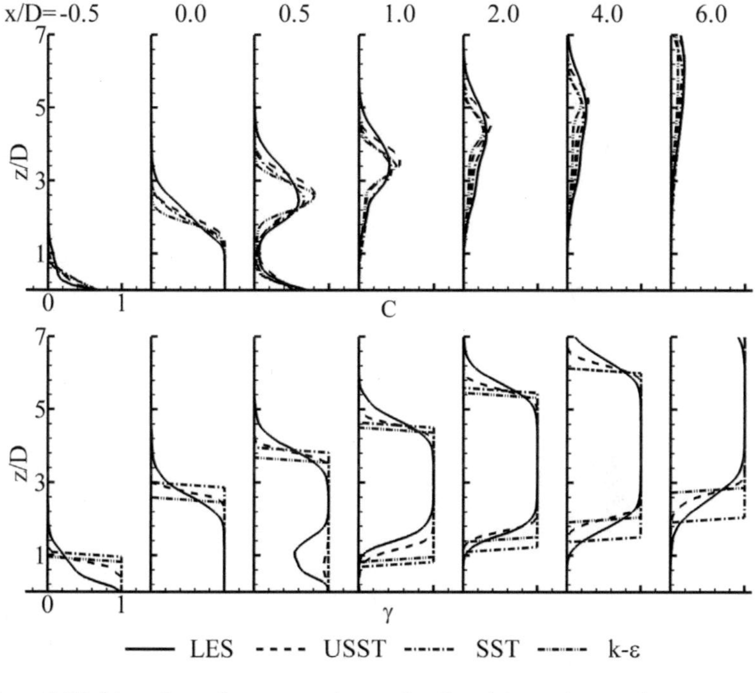

Figure 7.32: Line plots of mean passive scalar C and intermittency factor γ at the symmetry plane, $y/D = 0$, Case A.

Figure 7.33 shows two dimensional plots of the passive scalar C and the intermittency factor γ for the measurements and simulations at a horizontal plane, z/D = 1.5. Unfortunately experimental data about the intermittency factor γ is not available. The agreement of C predicted by the LES is very

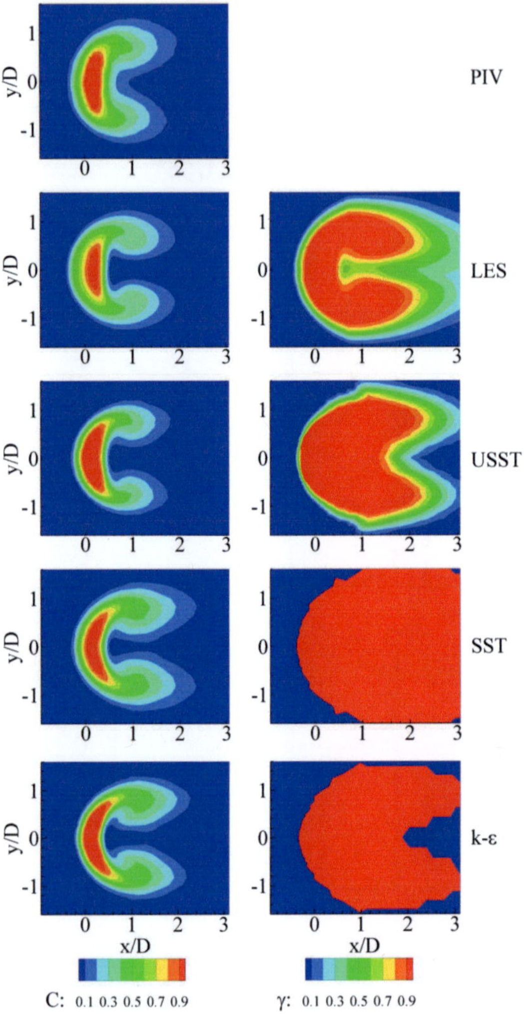

Figure 7.33: Two dimensional maps of mean passive scalar C and intermittency factor γ, $z/D = 1.5$, Case A. From top to bottom: LIF measurements, LES, unsteady SST, steady state SST and steady state k-ε simulations.

good, in both shape and level. The inner side of the kidney shape fits very well to the experimental data. Considering the unsteady SST simulation, the gradients of the scalar are steeper than in the LES, which makes the agreement with the measurements slightly worse. The kidney shape of the jet agrees well with the experimental data. One part of the differences in the prediction of C can be explained by the contours of γ. The LES is able to resolve a larger fraction of the large-scale coherent structures than the USST, which results in γ contours that are distributed over a larger area than the contours of the USST simulation; in other words, the LES predicts less steep gradients of γ than the USST, which could also be seen in the profiles in Figure 7.32. The agreement between the predictions of the LES and the USST is worse exactly in the regions with poor agreement of γ. For the steady state RANS simulations, the agreement of the predicted-to-measured scalar field C is not as favorable as was shown for the velocity fields. The core flow is reasonably well predicted by both simulations, however with a more elongated kidney shape. The RANS simulations have more difficulty predicting the borders of the jet, with the lips downstream being more elongated. This difficulty is expected, as the steady state RANS simulations do not resolve coherent structures and the influence of the intermittency is more pronounced in this region. In contrast to the results of the LES and USST, the RANS simulations predict γ contours with a sharp change between the jet and the ambient fluid.

The turbulent mixing can be analyzed in more detail using the line plots at the symmetry plane ($y/D = 0$) of the mean passive scalar C, the intermittency factor γ, the root mean square (rms) value of the passive scalar c_{rms} and the $\overline{u'c'}$ component of the specific Reynolds flux vector depicted in Figure 7.34 for three different z/D positions (1.5, 3 and 4.5). The C profiles show more pronounced differences than the mean velocity depicted in Figure 7.31 for the same locations. The LES simulation shows good agreement with the measurements, with slightly underpredicted peak scalar values at $z/D = 3$ and 4.5. The C profiles of the USST simulation have a good agreement, although not as good as the agreement of the mean velocity in Figure 7.31. The position of the jet agrees well, however the profiles of the simulations are slightly narrower than the measurements. The LES shows less steep gradients of γ than the USST, which is particularly evident at $z/D = 4.5$. Comparing the γ and C profiles, it can be seen that the broader C distribution in the case of the LES correlates very well with the also broader γ distribution. Comparing the steady state RANS simulations, the C profiles predicted by the SST simulation show good agreement with the measurements at $z/D = 1.5$, while at $z/D = 3$ and 4.5

it had problems predicting the jet borders correctly. Following the previous results, the simulation using the k-ε model has predicted an even narrower jet in all positions, which is expected, as the k-ε model is known to underpredict the mixing. The LES predicts c_{rms} in good agreement especially at $z/D = 1.5$,

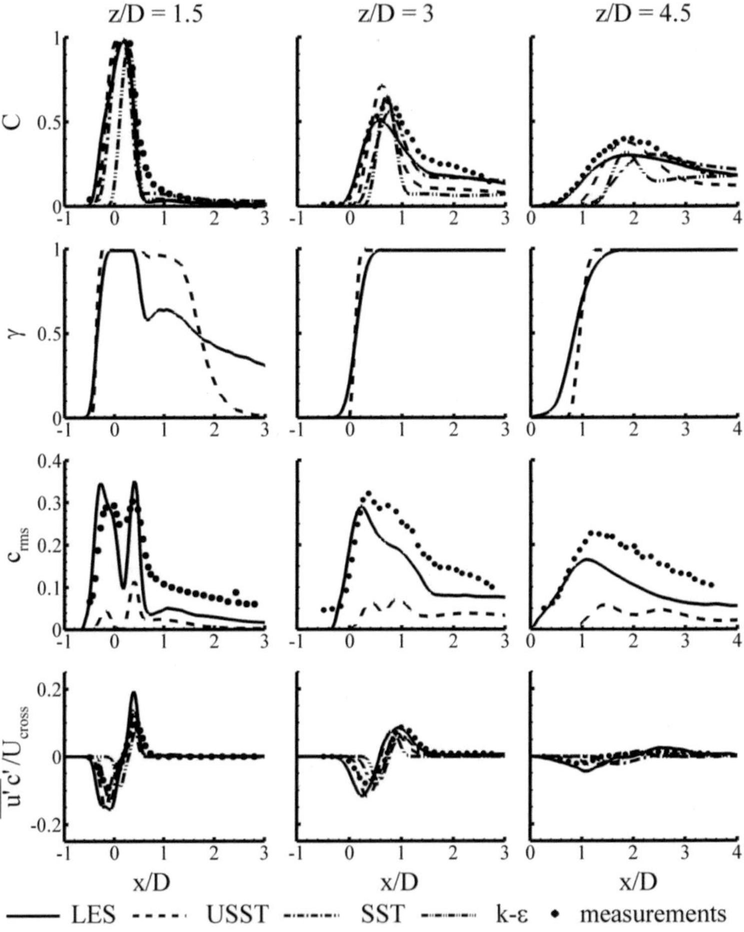

Figure 7.34: Line plots of mean passive scalar C, intermittency factor γ, passive scalar rms c_{rms} and specific Reynolds flux component $\overline{u'c'}/U_{cross}$ at the symmetry plane, $y/D = 0$, Case A.

with the two peaks at the jet boundaries being clearly discerned. However, the overall level is lower than in the measurements, despite the fact that the Reynolds stresses and fluxes are well represented. The c_{rms} levels of the USST simulation are clearly too low. The reason is that for the evaluation of c_{rms} only the unsteadiness of the flow was considered, as the RANS turbulence models do not predict c_{rms} directly. The RANS turbulence models do not predict c_{rms} directly, so no results are shown. The LES slightly overpredicts $\overline{u'c'}$ at $z/D = 1.5$, while for the other two locations the agreement is good. For the other simulations the agreement of $\overline{u'c'}$ is good, however the location of the peaks at $z/D = 3$ and 4.5 are slightly moved in the downstream direction in the steady state simulations. It should be noted that the measurements under these high turbulent conditions are particularly challenging, and always contain a certain amount of uncertainty. The evaluation of cross-correlations like $\overline{u'c'}$ is particularly sensitive, since they are an ensemble of various variables, which make the uncertainties being even more pronounced.

The comparison with the experimental data has shown that while the mean velocity is predicted well by all simulations, the prediction of the mean passive scalar C depends heavily on the level of complexity of the simulation. The results show that the agreement of C is very good using the LES, and degrades significantly when moving to USST, SST and k-ε simulations, in this order. This different behavior is attributed to the different way that the velocity and the passive scalar react to the presence of the jet. When portions of ambient fluid are entrained in the jet flow, they are accelerated before effectively meeting the jet flow, leading to velocity fluctuations. These fluctuations, that are not turbulent in nature, are induced by the pressure fluctuations associated with the turbulence [63].

Figure 7.35 shows a planar cut along the symmetry plane of a single time step of the LES of the Case A. The jet flow region is marked with gray color, while the color of the ambient fluid is white. The boundary between the jet flow and the ambient fluid is marked by the isocontour of $C = 0.01$, the same definition used to calculate the intermittency function. The boundary is highly convoluted, which is a result of the various coherent structures of the flow. A small region is shown in more detail, with arrows representing the local velocity vector. In the top right hand corner of the detail, a region of ambient fluid (white) is being entrained by the jet flow (gray). The entrained region is characterized by the velocity vectors, which have a component in the vertical direction. In contrast, in other areas away from the entrained region, the velocity vectors of the ambient fluid have components mainly in the horizontal direction. Although the difference in the velocity magnitude is not very pronounced, the velocity vectors of the ambient fluid in the entrained region show that the fluid has been accelerated in the vertical direction.

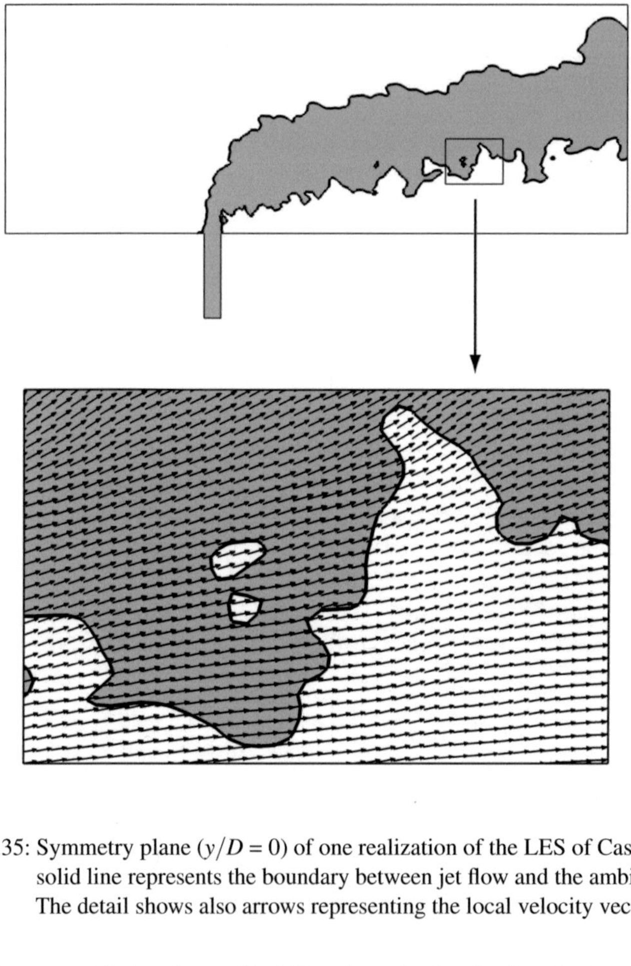

Figure 7.35: Symmetry plane ($y/D = 0$) of one realization of the LES of Case A. The solid line represents the boundary between jet flow and the ambient fluid. The detail shows also arrows representing the local velocity vector.

It corroborates with the theory [1, 93] and results for the free jet (see Section 6.3.5). As the fluid is accelerated before being effectively entrained, the velocity variation across the superlayer is smaller than the variation of the scalar. The analysis in Section 7.3.1 support this result, showing that the passive scalar flow rate across the superlayer is greater than the dimensionless velocity flow rate, which indicates that the passive scalar gradient is greater than the velocity gradient across the superlayer.

The conditional averages calculated with the LES can be used to analyze further the different behavior of the velocity and scalar fields. Figure 7.36 shows the profiles of the passive scalar C along with the mean values conditioned on the jet flow C_j and on the crossflow C_c. The intermittency factor γ is also plotted. As $C = 0$ in the crossflow, the value of C_c is zero in all profiles. In the jet core region, where $\gamma = 1$, the value of C is equal to C_j. In the lee side of the jet, even with decreasing value of γ, the values of C and C_j remain almost identical. This correlation is the same found in steady state simulations, and indicates a limited influence of the coherent structures in this region. On the upstream side of the jet the difference between C and C_j becomes evident, with the values of C_j reaching values around 0.1 and then suddenly dropping to zero. This indicates that, in this region, the mean passive scalar value diminishes as a consequence of the diminishing fraction of time that jet fluid flows there, however the value of the passive scalar inside the jet remains more constant toward the jet edge than the mean profiles suggest. This effect is caused by the large-scale coherent structures present in the flow, as suggested by Dahm and Dimotakis [19]. The turbulence inside large-scale structures is high, which leads to fairly homogeneous passive scalar value inside them. The value of C_j at the jet borders reflects this fairly homogeneous distribution.

Figure 7.37 shows line plots of the mean velocity component U along with the mean values conditioned on the jet flow U_j and on the crossflow U_c and the intermittency factor γ. In the lee side of the jet the three mean values U, U_j and U_c are closely correlated, with negligible influence of coherent structures. The value of U_c is different from the expected value for the crossflow ($U_c/U_{cross} \approx 1$), indicating a strong influence of the jet flow in this region. In the upstream side, U and U_c tend to the expected value of unity as γ tends to zero. The value of U_j deviates only slightly from the mean value U until dropping suddenly to zero. For the passive scalar, the discrepancy between C_j and C is much more evident in the region $0<\gamma<1$, where the influence of the intermittency is more relevant.

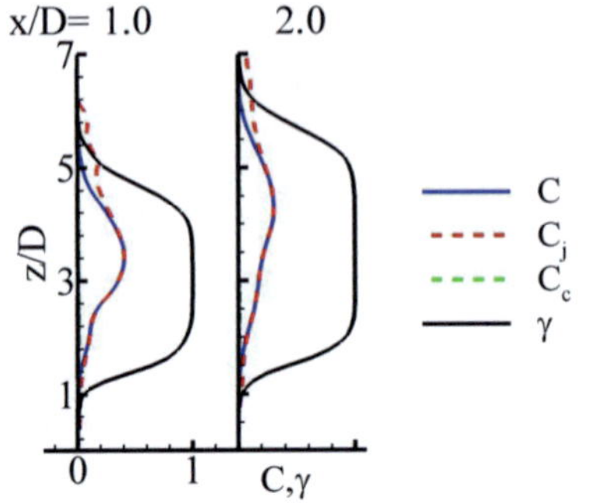

Figure 7.36: Line plots of mean passive scalar C, mean passive scalar conditioned on the jet flow C_j, conditioned on the crossflow C_c and the intermittency factor γ at the symmetry plane, $y/D = 0$, LES of Case A.

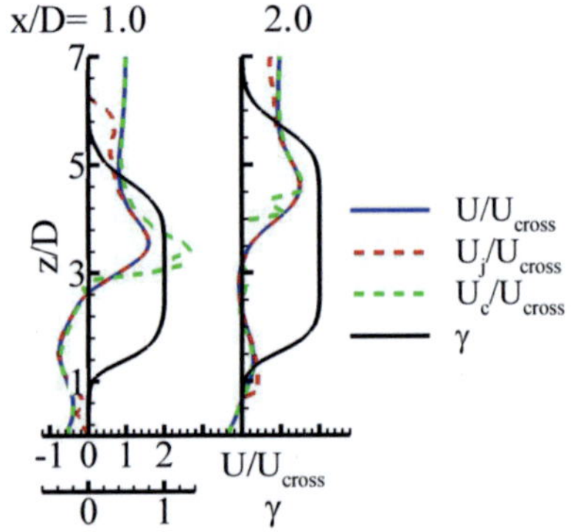

Figure 7.37: Line plots of mean velocity component U/U_{cross}, mean velocity component conditioned on the jet flow U_j/U_{cross}, conditioned on the crossflow U_c/U_{cross} and the intermittency factor γ at the symmetry plane, $y/D = 0$, LES of Case A.

In contrast, the discrepancy between U_j and U is less evident and limited to the jet edges where γ is close to zero. It is interesting to note that, in some regions of the jet core ($\gamma = 1$), the velocity conditioned on the crossflow U_c assumes higher values than in the pure crossflow at $x/D = 1$. It is argued that the reason for that phenomenon is the acceleration experienced by the ambient fluid that is entrained by the jet flow. This acceleration is induced by the pressure fluctuations associated with the jet flow [63].

It can be learned from Figures 7.36 and 7.37 that the intermittency has different effects on the velocity and scalar fields. It corroborates with the comparison of the measurements and the results of the simulations, which indicated good agreement of the mean velocity using all simulations, while for the passive scalar only the results of the LES and the USST, that are able to resolve the intermittency, had good agreement with the measurements.

7.3.5 Impact of different velocity ratios

After the extensive comparison of the simulations with the measurements of Case A, in this section the results of the simulations are compared to measurements of Cases B, C and D. The major difference between them is the velocity ratio, which increases from R=2.36 for Case B to R=4.49 for Case C and R=6.62 for Case D. Figure 7.38 show isosurfaces of passive scalar $C = 0.01$ for the LES of the three cases; the strong influence of the velocity ratio over the jet penetration depth can be clearly seen. As the measurements of Case A, these were also performed in the Engler-Bunte-Institute, Division of Combustion Technology (EBI-VBT) using the simultaneous PIV and LIF technique. All measurements are at the symmetry plane ($y/D = 0$), allowing the comparison of mean velocity and passive scalar and their fluctuations.

The results of the following simulations for each boundary condition are presented: a large eddy simulation (LES) with the standard Smagorinsky subgrid-scale model; an unsteady RANS simulation using the SST turbulence model (USST) and steady state RANS simulations using the SST turbulence model. The coarse grid was used in all simulations, in contrast to the previous section where the LES used the fine grid. It was shown in Section 7.2, especially Figure 7.4, that the effect of the grid resolution on the mean velocity was very limited. Although the effect on the turbulence is not negligible, the following analysis has shown that the results show the same trends and lead to the same conclusions as the simulations using the fine grid.

Case B

The velocity ratio of Case B is 2.36, which generates a jet that develops close to the bottom wall of the channel (see Figure 7.38). Figures 7.39 and 7.40 show line plots of mean velocity components U and W, specific Reynolds stress components $\overline{u'u'}$ and $\overline{w'w'}$, mean passive scalar C and passive scalar rms c_{rms} at the symmetry plane ($y/D = 0$). The penetration depth is predicted well by all simulations; the U profiles almost collapse. The same can be said from the W profiles, with a small deviation between the measurements and the simulations at $x/D = 0$ in the region below $z/D = 1$. As the W velocity component has its maximum at the core region of pipe flow, it is expected to decrease with increasing z/D as the pipe flow enters the channel, as has been seen in the measurements of Case A and is also predicted by the simulations. The measurements show a different behavior, with the jet velocity increasing when flowing into the domain until $z/D = 0.9$ and then decreasing. It is argued that it is due to some inconsistency in the measurements.

Following the trend in Case A, the $\overline{u'u'}$ and $\overline{w'w'}$ fields are substantially different between the simulations. The LES has a good agreement with the measurements, in both shape and level, excluding the region at $x/D = 0$ and below $z/D = 1$ discussed above. The USST simulation predicted lower levels of both Reynolds stress components, and the levels of the SST simulation are even lower. This deviation agrees well with the different level of description of the large-scale coherent structures by the different simulations: while the LES is able to resolve a large fraction of the coherent structures, the USST can resolve only a limited fraction of them and the SST resolve no coherent structures at all. A more detailed discussion can be seen in the results of Case A in Figure 7.27, and are also valid for the Case B in question.

The differences of the profiles of mean passive scalar C are more pronounced than the profiles of the mean velocity. The agreement of the LES is very good. Both RANS simulations, on the other hand, have predicted a distribution that is narrower as the measurements show. The passive scalar rms c_{rms} simulated by the LES is in good agreement with the measurements, with a slightly underpredicted peak value from $x/D = 1$, which can be related to use of the coarse grid. The USST predictions of c_{rms} consider only the unsteadiness of the flow, because the RANS models do not calculate c_{rms} directly. The results of the USST show levels of c_{rms} that are much lower than the measurements.

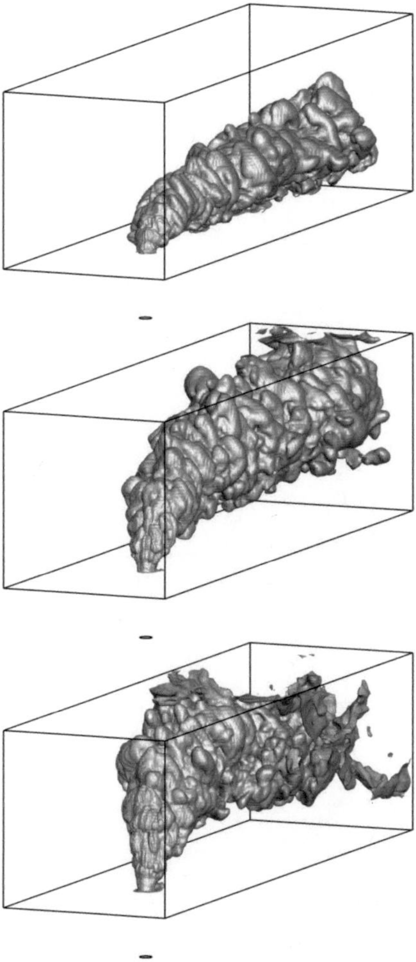

Case B, R = 2.36

Case C, R = 4.49

Case D, R = 6.62

Figure 7.38: Isosurface of passive scalar for Cases B, C and D.

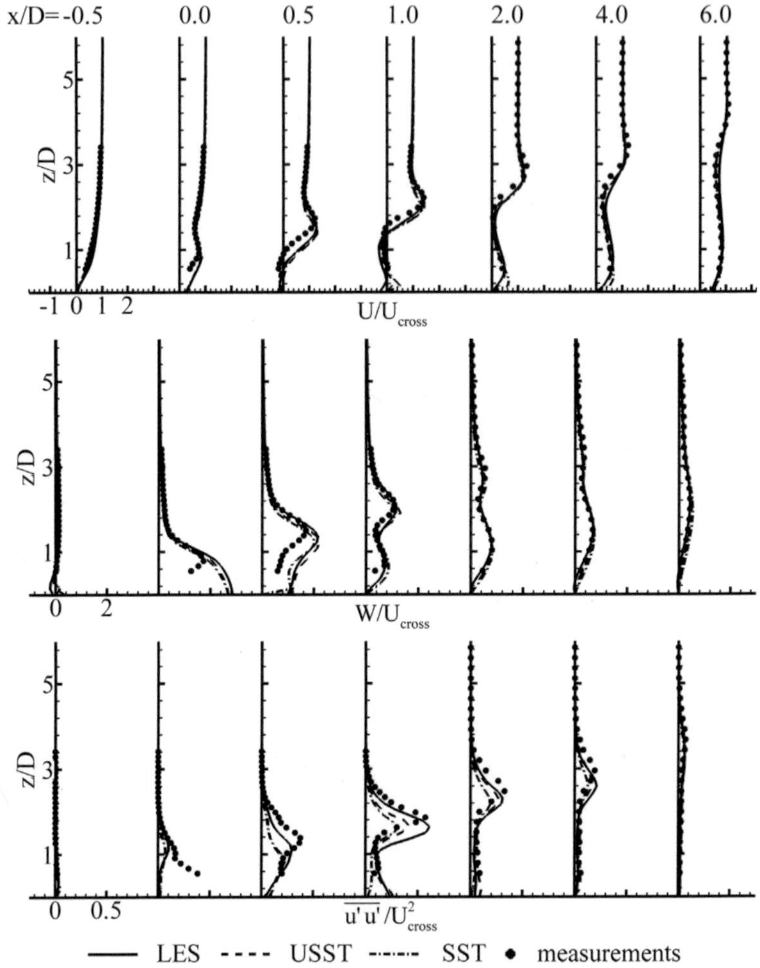

Figure 7.39: Line plots of mean velocity components U/U_{cross} and W/U_{cross} and the specific Reynolds stress component $\overline{u'u'}/U_{cross}^2$ at the symmetry plane, $y/D = 0$, Case B.

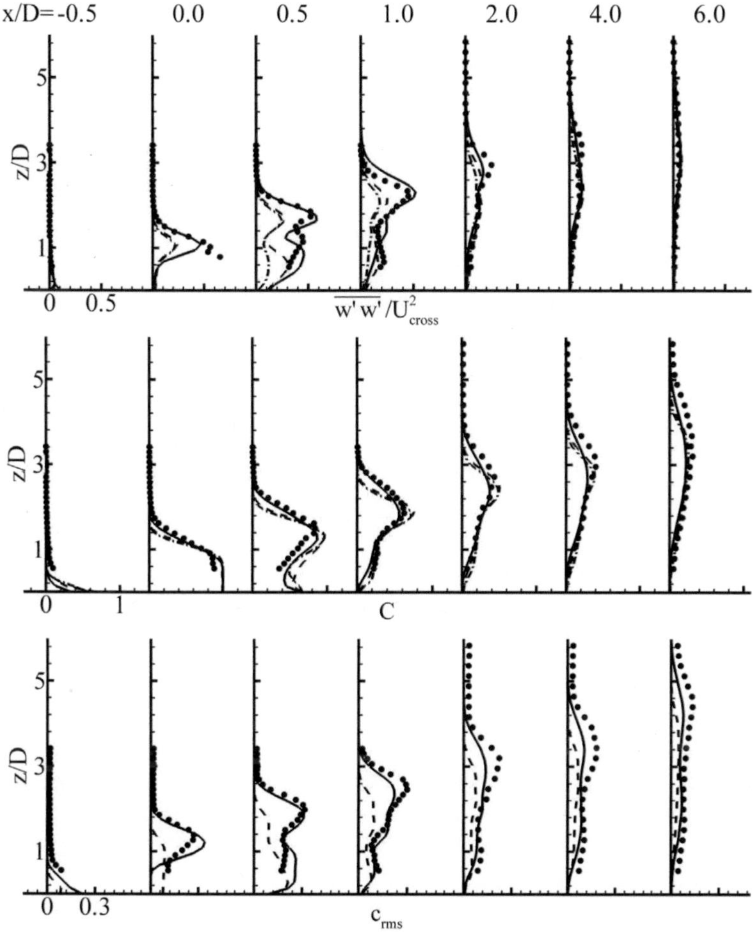

Figure 7.40: Line plots of specific Reynolds stress component $\overline{w'w'}/U^2_{cross}$, passive scalar C, passive scalar rms c_{rms} at the symmetry plane, $y/D = 0$, Case B.

Case C

The Case C has a velocity ratio of 4.49, which is close to the value of 4.15 from Case A. The jet flows approximately in the middle of the channel, which allow the undisturbed development of large-scale coherent structures (see Figure 7.38).

Figures 7.41 and 7.42 show line plots of the same variables as in Figures 7.39 and 7.40. As for Cases A and B, all simulations show good agreement with the measurements for the mean velocity components U and W. The penetration depth is also in good agreement, especially the LES.

The predictions of $\overline{u'u'}$ and $\overline{w'w'}$ follow the same trend seen in Cases A and B, with the LES having an overall higher fluctuation level than the USST and SST simulations. The agreement of $\overline{u'u'}$ predicted by the LES is not as good as seen in Case A, which can be attributed to the coarser grid used in the simulation of Case C. For $\overline{w'w'}$ the agreement of the LES with the measurements is better, with only a region at $x/D = 0.5$ and below $z/D = 2$ with a significant deviation. The fluctuation level of the USST and SST simulations is lower than the measurements and the LES. As discussed previously, this deviation is attributed to the different level of description of the coherent structures by the different simulations.

The mean passive scalar C of the LES shows a very good agreement with the measurements. Following the trend of Cases A and B, the RANS simulations have predicted a narrower distribution than the measurements show. The agreement of c_{rms} simulated by the LES is good, with a slightly underpredicted level in the region below $z/D = 4$ downstream of $x/D = 0.5$. As the USST predictions of c_{rms} consider only the unsteadiness of the flow, the profiles show much lower levels than the measurements.

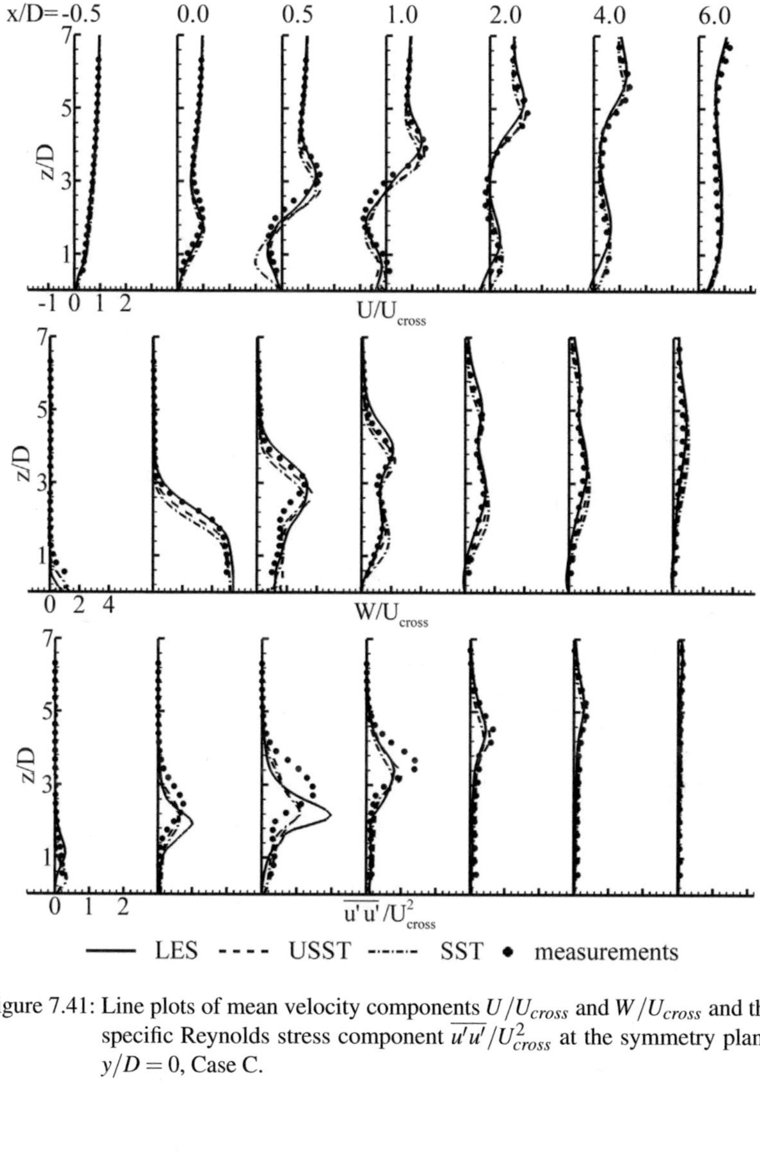

Figure 7.41: Line plots of mean velocity components U/U_{cross} and W/U_{cross} and the specific Reynolds stress component $\overline{u'u'}/U_{cross}^2$ at the symmetry plane, $y/D = 0$, Case C.

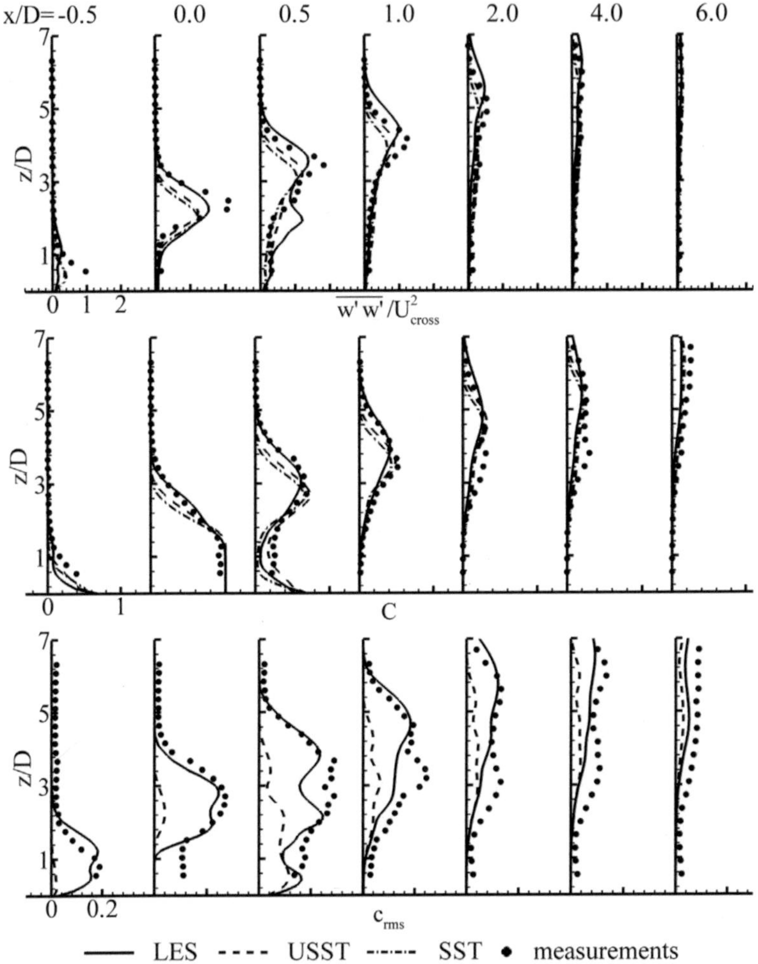

Figure 7.42: Line plots of specific Reynolds stress component $\overline{w'w'}/U^2_{cross}$, passive scalar C, passive scalar rms c_{rms} at the symmetry plane, $y/D = 0$, Case C.

Case D

The velocity ratio of Case D is the highest in this study with the value R=6.62, which leads the jet to reach the upper wall of the channel at a point near $x/D = 10$ (see Figure 7.38). This fact has only a limited influence at the results presented below, that span between x/D = -0.5 and 6. It has, however, an adverse impact on the development of large-scale coherent structures, that are deformed and eventually destroyed when reaching the upper wall.

Line plots of the same variables as in Figures 7.39 and 7.40 can be seen in Figures 7.43 and 7.44. The agreement of the mean velocity components U and W is good for all simulations, including the penetration depth. The deviation is slightly more pronounced only in the region at $x/D = 0.5$ and below $z/D = 4$, which is characterized by an intense recirculating flow. The recirculation region poses a challenge to both measurements and simulations.

The comparison of $\overline{u'u'}$ and $\overline{w'w'}$ shows a different behavior as the results for Cases A, B and C. The three simulations predicted levels of $\overline{u'u'}$ in good agreement with each other and to the measurements. The difference between the simulations became more visible for $\overline{w'w'}$, however the agreement is still good. Only the LES had more problems in the region at $x/D = 0.5$ and below $z/D = 4$ discussed above. As the higher velocity ratio of Case D leads the jet to reach the upper wall of the channel, the development of the large-scale coherent structures of the flow is adversely affected. The influence of the coherent structures is consequently less pronounced, which leads in turn to a less pronounced difference between the unsteady and steady state simulations. It should also be noted that the coarse grid has been used for the LES, which is not able to fully resolve the steep gradients of the flow, which can affect negatively the agreement with the measurements.

The turbulent mixing, represented by the mean passive scalar C, follow the trend seen for the other velocity ratios and show a clear difference between the simulations. The LES has a good agreement, while both RANS simulations have predicted a narrower distribution as the measurements show. The agreement of the LES profiles of passive scalar rms c_{rms} is satisfactory, with larger deviations between x/D = -0.5 and 0.5. The USST predictions of c_{rms} are much lower than the measurements, which is a consequence of only considering the unsteadiness of the flow to evaluate c_{rms}.

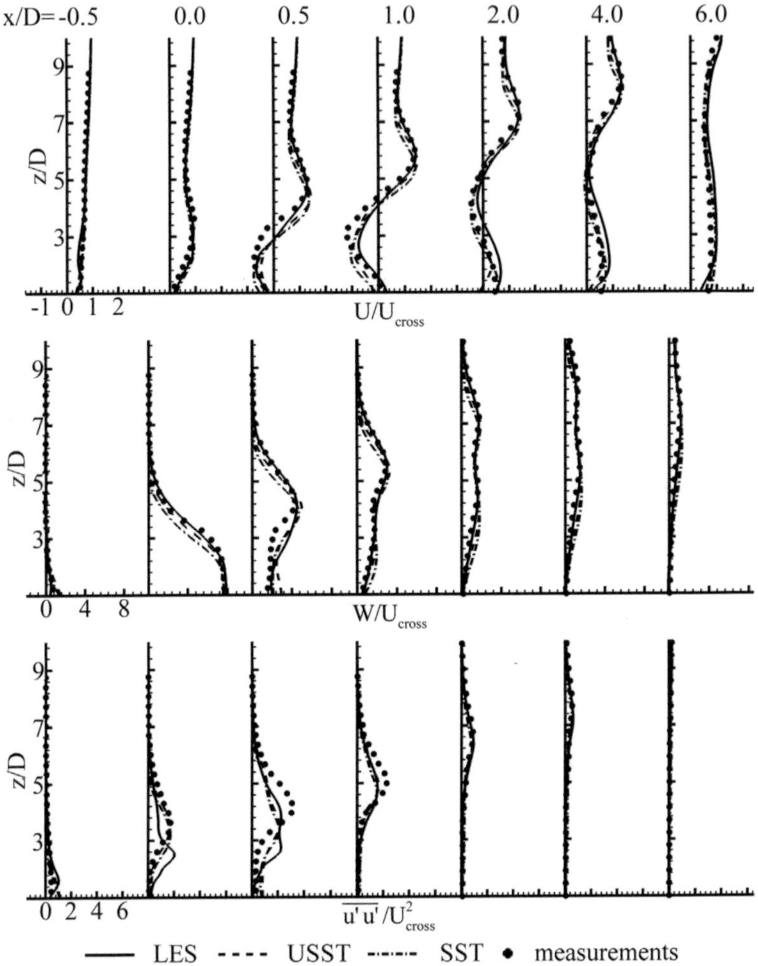

Figure 7.43: Line plots of mean velocity components U/U_{cross} and W/U_{cross} and the specific Reynolds stress component $\overline{u'u'}/U_{cross}^2$ at the symmetry plane, $y/D = 0$, Case D.

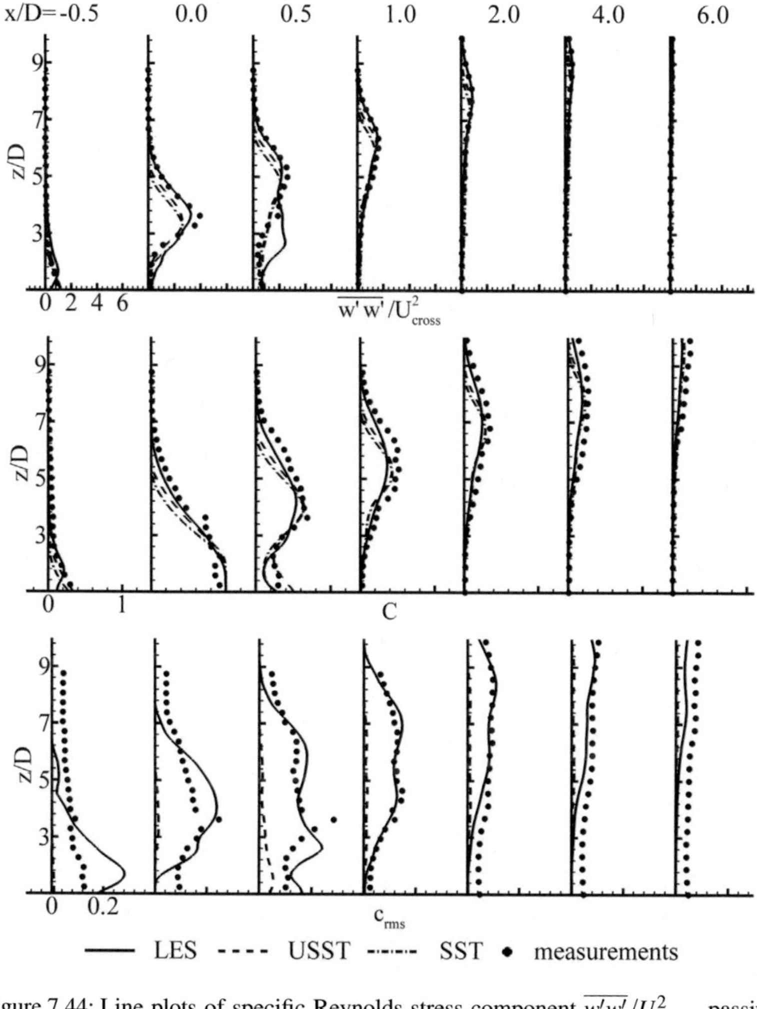

Figure 7.44: Line plots of specific Reynolds stress component $\overline{w'w'}/U_{cross}^2$, passive scalar C, passive scalar rms c_{rms} at the symmetry plane, $y/D = 0$, Case D.

An inconsistency should be noted in the measurements of c_{rms}. In the regions where the mean concentration C is zero there should be no concentration fluctuation and consequently c_{rms} should be zero; it is a consequence of the value of the passive scalar being bounded ($0 < c < 1$). In contrast, the measurements show a positive value of c_{rms} around 0.04 in the regions where $C = 0$.

7.3.6 Correlation between momentum and passive scalar turbulent diffusivity

Within the RANS framework introduced in Section 5.2, there are two quantities that describe the turbulent diffusivity of momentum and passive scalar: the eddy viscosity ν_t and the turbulent diffusivity D_t, respectively. The eddy viscosity is the factor of proportionality between the specific Reynolds stress tensor and the mean strain rate tensor S_{ij} through the Boussinesq approximation, Equation (5.4), repeated here for convenience

$$-\overline{u_i'u_j'} = 2\nu_t S_{ij} - \frac{2}{3}\rho k \delta_{ij}. \tag{7.5}$$

In the same way, the turbulent diffusivity is the factor of proportionality between the specific Reynolds flux vector and the mean scalar gradient through Equation (5.29), also repeated here

$$-\overline{u_i'c'} = D_t \frac{\partial C}{\partial x_i}. \tag{7.6}$$

The ratio between the two diffusivities is called turbulent Schmidt number $\sigma_t = \nu_t / D_t$.

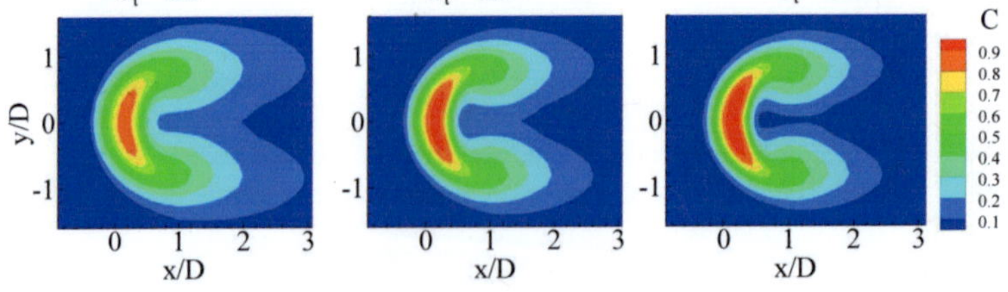

Figure 7.45: Two dimensional maps of mean passive scalar C, $z/D = 1.5$, Case A, for simulations using the SST turbulence model and turbulent Schmidt numbers σ_t of 0.3, 0.5 and 0.7.

One approach often used to enhance the mixing simulation is to decrease the standard value of the turbulent Schmidt number σ_t from 0.9 and consequently scale the Reynolds flux vector with it. In Figure 7.45 the results from three simulations using the SST turbulence model can be seen, with σ_t of 0.3, 0.5 and 0.7. The passive scalar field of the simulation using $\sigma_t = 0.3$ does not have the steep gradients of the standard simulation using $\sigma_t = 0.9$; this is consistent with the increased turbulent mixing. The above changes to σ_t affected the whole field simultaneously. The mixing in the jet core is well represented using the value of 0.9 while the borders appear to be better predicted using a lower value. This finding is consistent with the fact that the intermittency is higher at the jet borders; the deviations are expected to be higher in this region.

It is tempting to apply the definitions of Equations (7.5) and (7.6) to the experimental data and to the LES simulation and calculate σ_t, as both evaluate the Reynolds stresses and fluxes along with the gradients of the mean quantities. However, the application of these definitions is not straightforward. Within the RANS framework the eddy viscosity is a scalar quantity, which depends on the turbulent kinetic energy k and a length scale based either on the dissipation rate ε or on the specific dissipation ω. The turbulent diffusivity and the turbulent Schmidt number are also scalar quantities. On the other hand, when evaluating the results of the measurements, it becomes clear that the eddy viscosity has a different value for each component of the Reynolds stress tensor, being in fact a tensor quantity. The same applies to the turbulent diffusivity and to the turbulent Schmidt number, which are in fact vector quantities. The comparison of these tensor and vector quantities with the scalar results of RANS models demands great care.

Despite the shortcomings, much can be learned when analyzing the results of the measurements and the LES. The two-dimensional experimental data allows the calculation of one pair of eddy viscosity and turbulent diffusivity. For example, at the symmetry plane ($y/D = 0$) the Reynolds stress component $\overline{u'w'}$ and the Reynolds flux component $\overline{u'c'}$, together with the gradients of the mean quantities, can be evaluated. Figure 7.46 shows, for the measurements and for the LES of Case A at the symmetry plane, two-dimensional maps of the component of turbulent Schmidt number in the x-direction $\sigma_{t,x}$, together with the axial evolution at three vertical positions of $\overline{u'w'}$, $\overline{u'c'}$ and $\sigma_{t,x}$. To improve the readability of the picture, the plots were limited to regions with values of C between 0.05 and 0.95 and with absolute values of $\partial C/\partial x$ above 0.04.

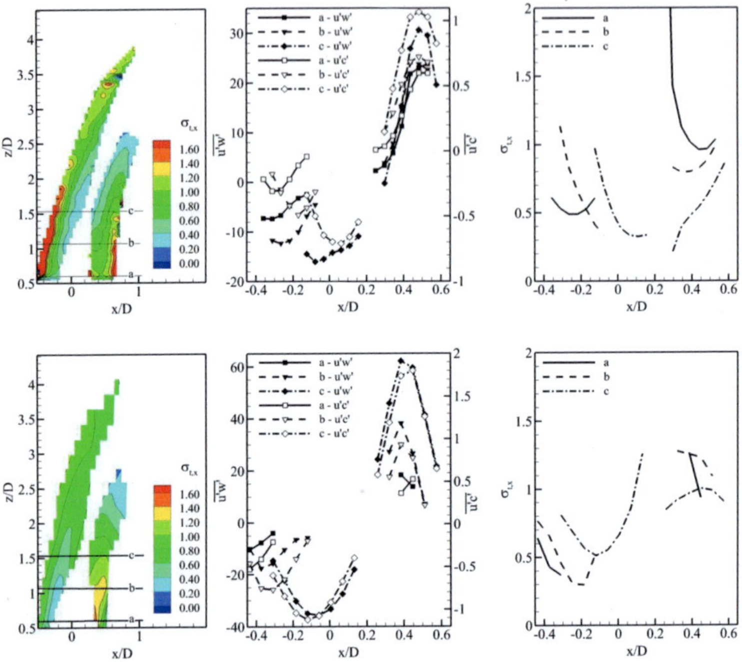

Figure 7.46: Two dimensional maps of turbulent Schmidt number $\sigma_{t,x}$ and line plots of the Reynolds stress component $\overline{u'w'}$, the Reynolds flux component $\overline{u'c'}$ and $\sigma_{t,x}$ at the symmetry plane, $y/D = 0$, for the measurements (top) and LES of Case A (bottom).

The major difficulty in comparing the turbulent Schmidt numbers from the measurements or the LES with the definition of the RANS framework is actually the fact that the turbulence is not the only source of fluctuation in the jet in crossflow. Equations (7.5) and (7.6) are derived assuming that turbulence is the only source of fluctuation in the flow, as all equations in RANS context. This is clearly not the case in the jet in crossflow, especially in regions where the coherent structures have already developed. The experimental results support this argument. At the jet root, near the nozzle, the coherent structures have not evolved sufficiently to affect the flow. In this region $\sigma_{t,x}$ assumes values between 0.3 and 1.3, which agrees well with the values recommended in the literature. Further downstream the influence of the coherent structures is more pronounced, and the values of $\sigma_{t,x}$ begin to fluctuate between very high and very low values, which is clearly indicated by the sudden appearance of red and blue regions in the contour plots from $z/D = 3$ and downstream. This finding leads to the conclusion that, in the presence of coherent structures, Equations (7.5) and (7.6) do not adequately represent the phenomena present there.

7.4 Concluding thoughts

The results of different simulations have been compared to high quality experimental data, for a series of boundary conditions. The LES has successfully predicted the various coherent structures of the flow, including the counter-rotating vortex pair (see Figure 7.12), the horseshoe vortex (see Figure 7.13) and the wake vortices, which shedding frequency has been analyzed in Section 7.3.2.

The agreement of the LES results with the measurements is very good for both the mean variables and the fluctuations. The unsteady SST simulation shows very good agreement with the mean variables. On the other hand, the fluctuation levels are systematically lower than the measurements, which can be attributed to the fraction of coherent structures that could not be resolved by the unsteady SST simulation. The agreement of the mean variables predicted by the steady state RANS simulations with the measurements is good, with the agreement of the mean velocity being systematically better than the mean passive scalar. The fluctuation levels predicted by the steady state simulations are significantly lower than the measurements, the reason being that these simulations do not resolve the coherent structures at all, neglecting an important source of unsteadiness.

A detailed analysis has shown that, in the flow regions dominated by coherent structures, the turbulent Schmidt number σ_t does not adequately represent the mixing phenomena present there.

8 High Strain Burner

Tools for simulating turbulent mixing with a high level of accuracy have been developed and tested in free jet (Chapter 6) and jet in crossflow (Chapter 7 configurations. In this chapter, these tools are employed in the simulation of a complex combustion system, in which the turbulent mixing plays a major role in the flame geometry and consequently in the whole system.

This chapter begins with a review of a short introduction about the combustion system being studied. It is followed by the experimental setup and the measurements. The chapter proceeds with the numerical setup and the results of the simulations. The comparison of the results of the simulations with the measurements closes this chapter.

8.1 Introduction

The focus of this study, which was financially supported by the Ministry of Research of Baden-Württemberg, Germany together with Siemens AG Germany within the special research initiative "Kraftwerke des 21. Jahrhunderts" ("Power Plants of the 21th Century"), is on the combustion zone generated by a combustible mixture injected into the crossflow stream of vitiated air in a two staged combustion system. The combustible mixture is composed of methane and air at different proportions, while the vitiated air stream is composed of exhaust gases containing a reduced percentage of oxygen compared to air. Figure 8.1 shows the layout of the experimental setup, which will be explained in more detail in the next section. The experiments have been conducted by Dr. Prathap Chockalingam, and the results have been partially published by Galeazzo *et al.* [35]. This system was given the name *High Strain Burner*, as the combustible mixture is injected with high velocities (and consequently high strain) into the vitiated air stream, causing the flame to burn lifted. In many practical burners in gas turbine combustors, lifted turbulent flames are employed. Significant efforts had been invested by many researchers to understand the stabilization of lifted non-premixed flames [90, 93, 71, 60].

Lifted turbulent flames generate significantly lower pollutant emissions than burner attached flames because of the mixing of the fuel with the surrounding oxidizer in the lift-off regime. A detailed review on stabilization and blow out mechanisms of turbulent jet flames was given by Pitts [90]. Peters [89] quoted that theory based on partial premixing predicts the lift-off height of a turbulent jet flame that matches closely with the experimental data. Lyons [71] in his review mentioned that theories based on partial premixing and edge flames are the foremost approaches considered for the determination of lift-off height of the diffusion flame. In a recent review, Lawn [60] had discussed the effect of coflow fluid on the stabilization of turbulent jet flames. The above mentioned reviews mainly discussed the stabilization mechanisms and the determination or estimation of lift-off height of the turbulent jet flames with or without coflow studied at atmospheric condition.

Results obtained with a similar system were already published by Prathap *et al.* [95, 96], with emphasis on the NO_X formation. An axially staged combustion system at elevated pressure conditions was studied experimentally and numerically. Pure methane was injected into the crossflow stream of vitiated air, and its effect on the NO_X formation in the secondary stage was measured. They reported that the mole fraction of NO_X formed in the secondary stage was lesser than 15 ppm for the investigated operating conditions. Although the steady state RANS simulations reported in references [95, 96] could reproduce qualitatively the non-similarity of the penetration depth with the pressure, which was observed experimentally, the agreement of the flame position was not quite satisfactory.

The main objective of the present study was to analyze a turbulent lifted premixed flame anchored in a crossflow of vitiated air in the second stage of a two staged combustor at elevated pressure conditions, in which the turbulent mixing plays a major role in the flame geometry. The chemiluminescence technique was used to image the secondary combustion zone experimentally. Unsteady large eddy simulations (LES) and steady state RANS simulations using the presumed JPDF turbulent reaction model were used to simulate the combustion in the secondary combustor.

8.2 Experiments

The layout of the experimental setup can be seen in Figure 8.1. The experimental setup has two stages. The first stage comprised a pulsation damper (not shown in the picture) and a primary combustor. The pulsation damper was used to ensure a uniform flow rate of combustion air in the air blast nozzle by eliminating the fluctuations in the incoming compressed air. The primary combustor was equipped with a low swirl air blast nozzle operating with kerosene, generating the vitiated air for the second stage.

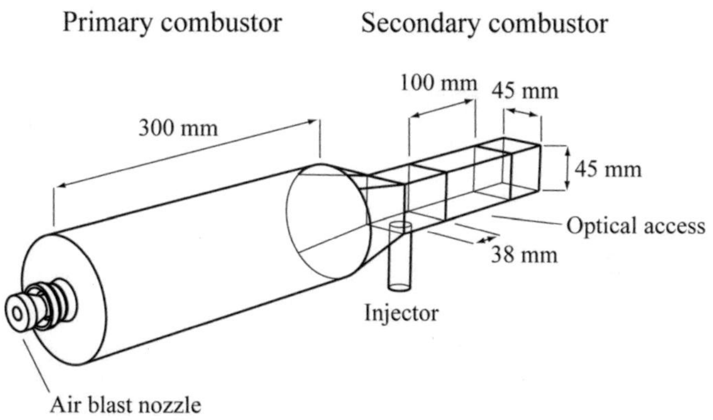

Figure 8.1: Layout of the experimental setup with dimensions. The flow direction is from left to right.

Then the secondary combustor followed the primary combustor as shown in Figure 8.1. The secondary combustor possessed optical access. It was quadratic in shape with inner dimensions of 45 x 45 mm. The size of the optical windows of the secondary combustor was 45 x 100 mm (41 x 90 mm after counting the wall interference effects). The secondary stage injector was a tubular burner with an inner diameter of 20 mm and a length of 327 mm. It was equipped with a ceramic porous disc mounted after the gas and air inlets in the injector to mix the air and methane flows. The injector was flush mounted in the secondary stage combustor. The distance between the axis of the injector and the leading edge of the optical windows was 38 mm. The optical rig could be operated up to 5 bar and with vitiated air temperatures up to 1800 K.

Monitoring the flame front itself is a difficult task. The OH radical is one important intermediate species in the oxidation of hydrocarbon fuels, being used as a marker for the flame front. In the chemiluminescence technique, a camera records the light emitted by the chemically excited OH, denoted OH*.

Unlike the OH molecule, which exists as an equilibrium product in regions of high temperature, the chemically excited OH* is short-lived and results from chemical reactions in regions with high heat release. Chemiluminescence imaging technique equipped with an ICCD (intensified charge-coupled device) camera was used to image the OH* emissions of the secondary stage flame. More details concerning the experimental setup can be found in Galeazzo *et al.* [35].

8.2.1 Operation procedure

The operating conditions of the experiments are summarized in Table 8.1. The air required for primary combustion was preheated up to 700 K and then supplied to the pulsation damper. The kerosene fuel was injected directly into the axis of the air blast nozzle using a hollow cone injector. To approach the desired operating conditions, the equivalence ratio of the primary combustor was close to 0.44. The temperature of the vitiated air was monitored using three thermocouples located at the bottom flange of the primary combustor. The pressure drop across the air blast nozzle and the absolute pressure in the combustion chamber was monitored using absolute pressure transducers. The mass flow rates of air and methane to the secondary stage tubular burner were measured using digital mass flow meters.

Initial temperature of primary combustion air	700 K
Pressure of primary combustor	2 bar
Adiabatic flame temperature in the primary stage	1820 K
Equivalence ratio of primary stage combustible mixture	about 0.44
Initial temperature of secondary stage mixture	300 K
Equivalence ratio of secondary stage combustible mixture	0.55 to 1.02

Table 8.1: Operating conditions.

To image the secondary combustion zone, chemiluminescence technique was used. It comprised of an ICCD camera mounted with a UV objective. The camera chip has a maximal resolution of 512 x 512 pixels, with a sampling rate of 3.5 Hz. The camera was positioned perpendicular to the injector. The light emitted by the OH* radicals has a maximum wavelength around 306.4 nm [58].

Once the desired operating conditions of the vitiated air in the secondary optical combustor were achieved, the premixed methane and air mixture was injected into the secondary combustor using the secondary stage injector. Then, the secondary stage combustion zone stabilized in the secondary combustor was imaged using the ICCD camera. The same procedure was repeated for all the investigated operating conditions.

8.2.2 Experimental results

Figure 8.2 shows the mean value of chemiluminescence intensity emitted by OH* radicals produced in the secondary stage combustion zone. The mean value was calculated from 200 images using a MATLAB code. As mentioned earlier, an ICCD camera was used for imaging the chemiluminescence emissions from OH* radicals. The resolution of the ICCD camera was 512 x 512 pixels, which corresponds to a physical distance of 120.3 x 120.3 mm. The field of optical access after subtracting the regions under the influence of wall effects (*i.e.*, 41 x 90 mm) is shown in Figure 8.2. At $\phi = 0.55$, the secondary stage combustion zone stabilized well away from the injector in the crossflow of vitiated air. With further increase in the equivalence ratio of the secondary stage combustible mixture to 0.77, the secondary stage combustion zone moves upstream. At stoichiometric condition, it stabilizes very close to the injector and the flame base is not visible in the figure as it is located in the non-optical access zone. The reason for this behavior is the turbulent burning velocity, defined as the speed at which the flame would propagate through a quiescent mixture of unburned reactants, which depends on the stoichiometry of the reactants, properties of the flow and turbulence. As the flow field remains the same for all conditions, the reason for the different stabilization points lies in the stoichiometry. The increase in the equivalence ratio from $\phi = 0.55$ to 1.02 corresponds to an increase in the burning velocity, as the equivalence ratio approaches the stoichiometric value. The flame stabilizes in a region where the turbulent burning velocity is able to sustain the incoming flow velocity, leading to flame stabilization [91]. The increased burning velocity leads the flame to stabilize in regions with higher flow velocity, which are closer to the injection point.

To estimate the approximate location of the secondary stage combustion zone of these images with respect to the axis of the injector, the first occurrence of maximum intensity of the corresponding operating condition was used. Following the mentioned procedure, the estimated axial locations of the secondary stage combustion zone were for $\phi = 0.55$ at 112.5 mm, for $\phi = 0.77$ at 74.3 mm and for $\phi = 1.02$ at 61.8 mm.

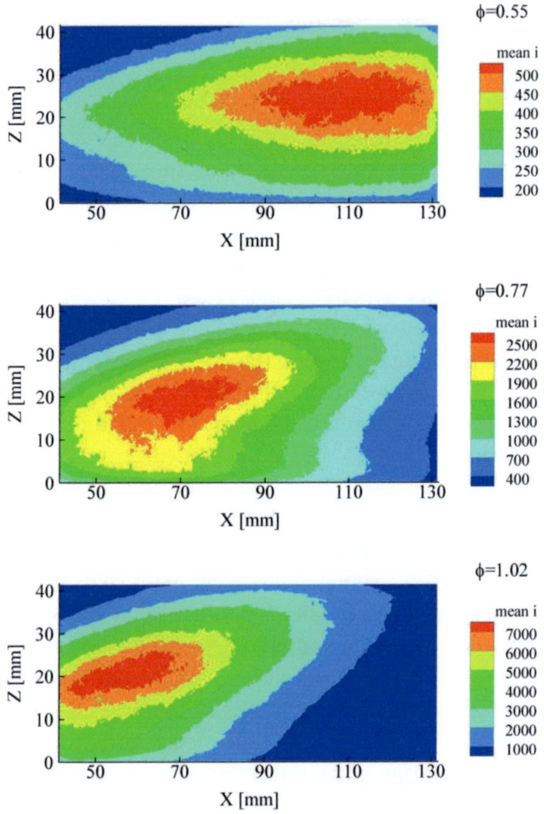

Figure 8.2: Mean (200 images) OH* chemiluminescence images of secondary stage combustion zone for the following equivalence ratios in the secondary stage: from top to bottom, (1) $\phi = 0.55$, (2) $\phi = 0.77$ and (3) $\phi = 1.02$.

8.3 Numerical setup

The effect of the change in the equivalence ratio of the secondary stage premixed methane-air mixture injected into the crossflow stream of vitiated air on the secondary stage combustion zone has been numerically studied. Figure 8.1 shows the investigated computational domain. It comprised of the primary and secondary combustors, with the same dimensions than the experimental rig. As the present focus was mainly on the flame stabilized in the secondary combustor and also to reduce the complexity of the simulation, the primary combustor was not modeled in detail. The vitiated air was introduced through the nozzle mounted in the primary combustor. The values of inlet conditions of the vitiated air were obtained from the measurements.

Now in the secondary combustor, the secondary stage premixed methane-air mixture was injected into the crossflow stream of vitiated air from the primary combustor, which leads to a stabilization of secondary stage flame. This configuration resembles a typical jet in crossflow, a flow configuration known by its features like large-scale coherent structures and recirculation regions, as seen in Chapter 7 and in the literature, *i.e.* [72, 31]. Galeazzo *et al.* [31] reported that large eddy simulation (LES) predicts the flow configuration in a jet in crossflow with very high accuracy. In order to investigate this flow, which resembles a jet in crossflow configuration with additional chemical reactions, LES and steady state RANS simulations were used.

A solver developed using the CFD software package OpenFOAM [84] was used for the simulations. For the LES, the subgrid scale turbulence was modeled employing the Smagorinsky model, with the Smagorinsky constant set to 0.1. The turbulence in the RANS simulations was modeled using the k-ε model.

The turbulent fluctuations in the LES were generated at the inlets using a boundary condition based on the work of Klein *et al.* [53], described in Section 5.7.3. The flow through the primary combustor was modeled with 5% turbulent intensity and a length scale of 5 mm, approximately 1/6 of the nozzle diameter.

The non-reflecting boundary condition, described in Section 5.7.3, was used in the outlet, allowing the pressure waves to flow out of the computational domain.

The numerical grid used in the simulations is three-dimensional, as required by LES, with more than 2 million hexahedral elements. Figure 8.3 shows the numerical grid at the symmetry plane in the region near the methane injector. The time step was 0.5 microseconds long, resulting in a CFL number of about 0.3. Approximately 115 characteristic time steps, calculated using the injector nozzle diameter and the bulk velocity of the vitiated air in the secondary combustor $D/U_{\text{vitiated air}}$, were simulated. It means that the vitiated air has flown over 115 times through the nozzle diameter during the averaging time, which corresponds to 0.04 seconds and 80,000 time steps.

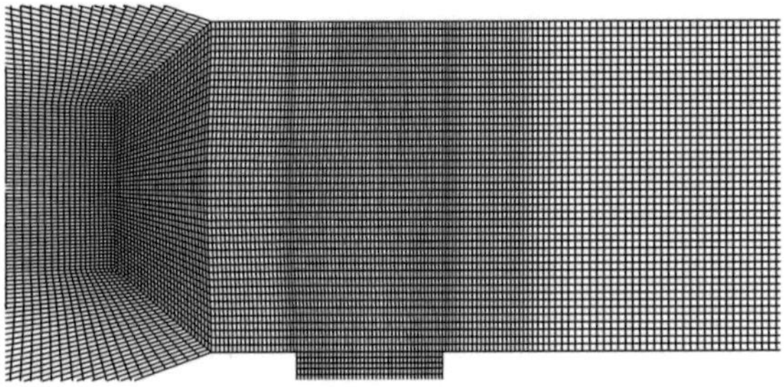

Figure 8.3: Numerical grid at the symmetry plane near the injector.

The description of the interaction between turbulent mixing and chemical reaction is a main challenge in the simulation of complex combustion systems. Prathap *et al.* [95, 96], analyzing a very similar system, compared the performance of two different modeling approaches: a combination of the eddy dissipation and the finite rate chemistry models and the presumed joint probability density function (JPDF) model. The results showed that the simulations were in closer agreement with the measurements when employing the presumed JPDF model. Consequently, the presumed JPDF model was used in the present simulations. The model is presented in detail in Section 5.6.

Two operating conditions, corresponding to the maximum and minimum equivalence ratios at 2 bar operating pressure were simulated. The boundary conditions are shown in Table 8.2.

Simulation	1	2
Absolute pressure (bar)	2	2
Equivalence ratio of secondary stage combustible mixture	0.55	1.02
Vitiated air temperature (K)	1696	1696
Mass flow of the vitiated air (kg/h)	169.5	171.5
Mass flow of combustible mixture (kg/h)	72.2	75.2
Mass composition of the vitiated air		
Y_{O_2}	0.1110	0.1119
Y_{CO_2}	0.1030	0.1028
Y_{H_2O}	0.0405	0.0403
Y_{N_2}	0.7455	0.7450

Table 8.2: Boundary conditions for the simulations.

8.4 Results and discussion

One of the characteristics observed in the LES is an acoustic resonance detected in the methane injector of the two staged combustion system under consideration. This effect, being essentially time-dependent, is not present in the steady state RANS simulations. New measurements have been planned to confirm the presence of this resonance effect in the methane injector, as the current measurements are not conclusive about it. Nevertheless, the resonance effect observed in the unsteady simulations was analyzed in detail using a computational domain that included only the secondary combustion chamber and the methane injector (see Figure 8.4). The equivalence ratio was set to 1.02 and a plug flow profile was applied to the vitiated air inlet. The other boundary conditions followed the values in Table 8.2.

In the injector, the methane-air mixture flows through a ceramic porous disc in order to enhance the mixing of the components, which has the side effect of damping the turbulence of the flow. A LES that could fully resolve the interaction of the ceramic porous disc with the pressure waves generated by resonance would require a computational grid describing each pore of the porous body, which is out of the scope of this study. Two different boundary conditions for the pressure were studied instead: fully reflecting and non-reflecting boundary conditions (BC), which represents the porous body as a solid wall and as an open inlet, respectively. The development of the pressure inside the methane injector is shown in Figure 8.5. The starting solution is the same, and after 0.04 s the solutions start to diverge from each other. The amplitude of the pressure waves is slightly higher using the fully reflecting BC than using the non-reflecting BC, which is a consequence of the pressure waves being reflected into the computational domain when using it. The frequency of the pressure waves, on the other hand, remains practically unaffected by the choice of BC. The average time between the pressure waves is 0.00137 s, resulting in a frequency of approximately 730 Hz.

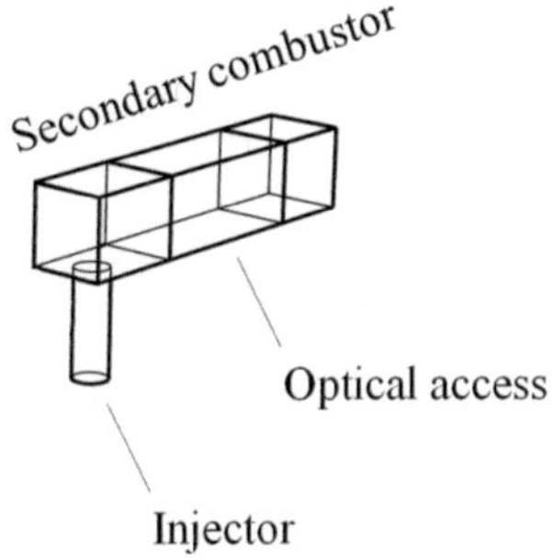

Figure 8.4: Computational domain for the study of the resonance in the methane injector, including only the secondary combustor and the methane injector.

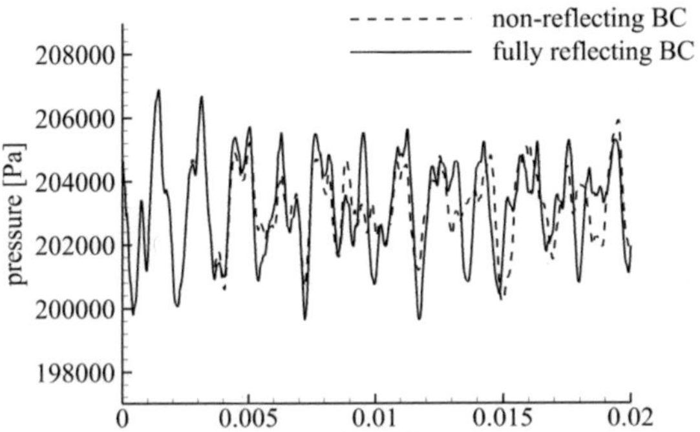

Figure 8.5: Pressure inside the methane injector over time. LES with fully reflecting and non-reflecting boundary conditions.

The resonance frequency of a closed tube can be approximated using the following equation:

$$f = \frac{na}{4(L + 0.4\,d)} \tag{8.1}$$

where n here is an odd number (1, 3, 5...), a is the speed of sound, L the length of the tube and d the diameter of the tube. Substituting the values for the present case and using $n = 3$, the resonance frequency is estimated to be 774 Hz, in very good agreement with the value of 730 Hz predicted by the simulations.

The effect of the resonance on the flow can be visualized in Figure 8.6, where the time evolution of the contours of mixture fraction makes the injector flow visible. The time elapsed between the frames is 0.002 s. The pressure waves promote a pulsation of the flow inside the methane injector, with considerable consequences to the mixing of the jet with the vitiated air flow. The choice of boundary conditions has a small impact in the mean mixture fraction, as depicted in Figure 8.7. The mean mixture fraction is almost the same using the fully reflecting or the non-reflecting boundary conditions. The impact of using RANS or LES on the mixing will be discussed in more detail using the results for the whole system in Figures 8.9 and 8.10.

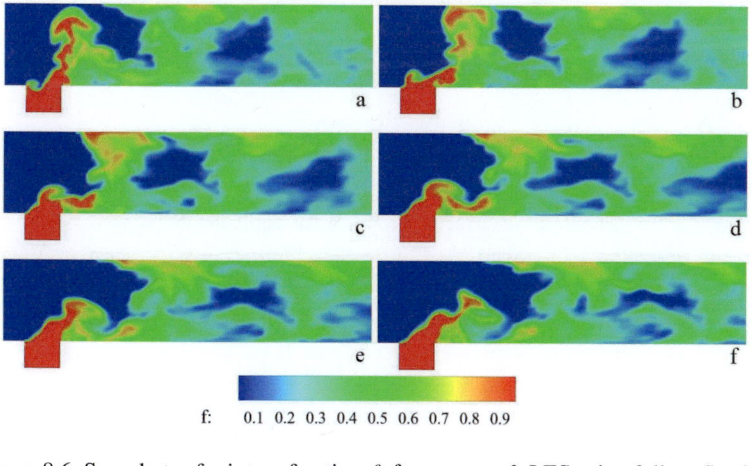

Figure 8.6: Snapshots of mixture fraction f, frames a to f. LES using fully reflecting boundary condition.

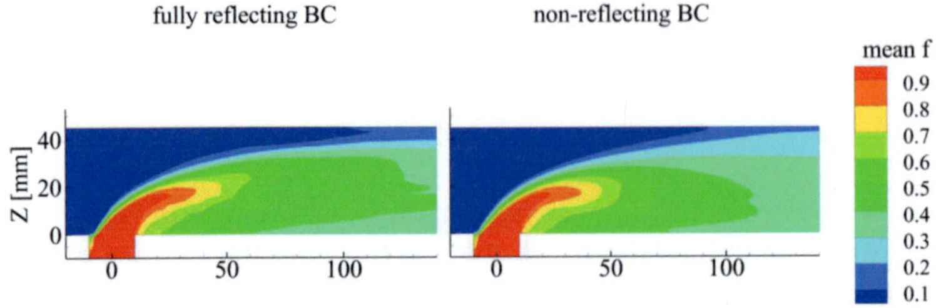

Figure 8.7: Contours of mean mixture fraction f. LES using the fully reflecting and the non-reflecting boundary conditions.

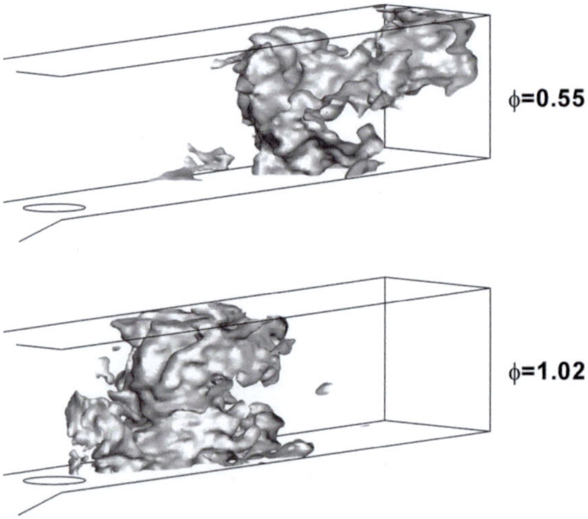

Figure 8.8: Snapshots of isosurface of $c = 0.5$ for equivalence ratios $\phi = 0.55$ and $\phi = 1.02$.

Now the focus will be shifted to the simulation of the whole system, with primary and secondary combustors (see Figure 8.1). The development of flame is described by the reaction progress variable c. Figure 8.8 shows snapshots of $c = 0.5$, which indicates the location of the flame front, for the LES with equivalence ratios 0.55 and 1.02. The flame fronts are highly distorted due to high turbulent intensities and large-scale coherent structures imparted by the jet in crossflow configuration. Figure 8.8 indicates that the flame at $\phi = 1.02$

burns lifted and stabilized in the immediate vicinity of the injector in the flow direction of vitiated air. However, at $\phi = 0.55$ the combustion zone stabilized further downstream and well away from the injector, which corroborates with the measurements (see Figure 8.2).

Figure 8.9 shows the contours of mean mixture fraction f, mean reaction progress variable c and mean temperature in the symmetry plane ($y = 0$) of the secondary combustor for the steady state RANS and LES simulations with $\phi = 0.55$. There is a very pronounced difference between the results of the RANS and the LES. While in the RANS simulations the jet develops closer to the bottom wall, in the LES the jet touches the upper wall in a short distance. The mixing is more intense in the LES, which can be seen observing the more homogeneous color distribution of mixture fraction. The cause of the large deviation between the RANS and the LES is the same as discussed in Chapter 7 for the jet in crossflow: the RANS simulations neglect an important source of unsteadiness, which has a direct impact in the description of the mixing. In the case of the jet in crossflow, the unsteadiness was caused by the coherent structures. In the combustion system being analyzed, besides neglecting the coherent structures, also the resonance effect in the methane injector cannot be reproduced by the steady state RANS simulations, with severe consequences for the mixing and consequently for the whole combustion system.

The large deviation between the RANS and the LES can also be seen in the contours of temperature, while the mean flame position, denoted by the reaction progress variable, do not deviate as much as the mixture fraction.

Figure 8.10 shows the same results as Figure 8.9, however with $\phi = 1.02$. Analyzing the mean mixture fraction f, the contours of lean and stoichiometric conditions should be nearly the same due to the presence of identical boundary conditions until $x = 30$ mm. For stoichiometric condition, at $x = 30$ mm, the combustion started in the secondary stage and the contours of mean mixture fraction between the stoichiometric and lean conditions start to deviate. Comparing the contour of f of the RANS simulations with $\phi = 0.55$ and $\phi = 1.02$ shows a very good agreement; the same can be said about the LES. On the other hand, when comparing the RANS and the LES for the same stoichiometry, the agreement is worse. One of the reasons is that the RANS simulations are not able to solve the coherent structures and the resonance effects on the flow. For example, there is a large deviation between the secondary combustion zone between the RANS and LES simulations for $\phi = 1.02$. While the combustion zone predicted by the LES starts near the injector, the RANS predicted a flame front with a very different shape. The flame with $\phi = 0.55$ stabilized further downstream, as expected, and its maximum temperature is lower than the maximum temperature obtained for $\phi = 1.02$.

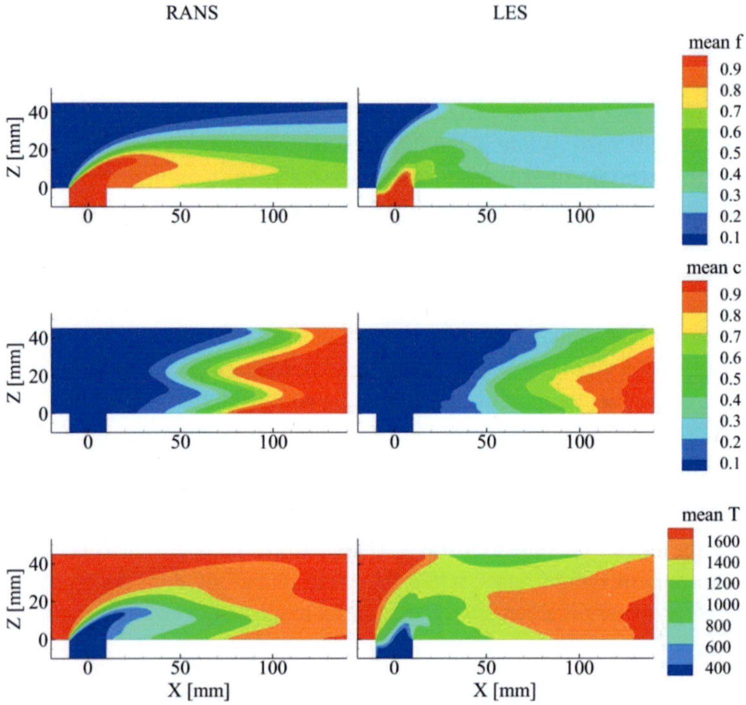

Figure 8.9: Contour of mean mixture fraction f; mean reaction progress variable c and mean temperature T at the symmetry plane, $y = 0$, for RANS and LES simulations with equivalence ratio $\phi = 0.55$.

It is important to mention here that the flame is not as thick as suggested by the contours of the time-mean reaction progress variable. In the case of stoichiometric condition, the value of reaction progress takes between 50 and 110 mm to develop from the unburned ($c = 0$) to the fully burned state ($c = 1$). This represents only the average position of the flame front; the time-resolved flame thickness is much smaller. The large flame thickness is related to the significant variation in the position of the flame caused by the turbulent flow.

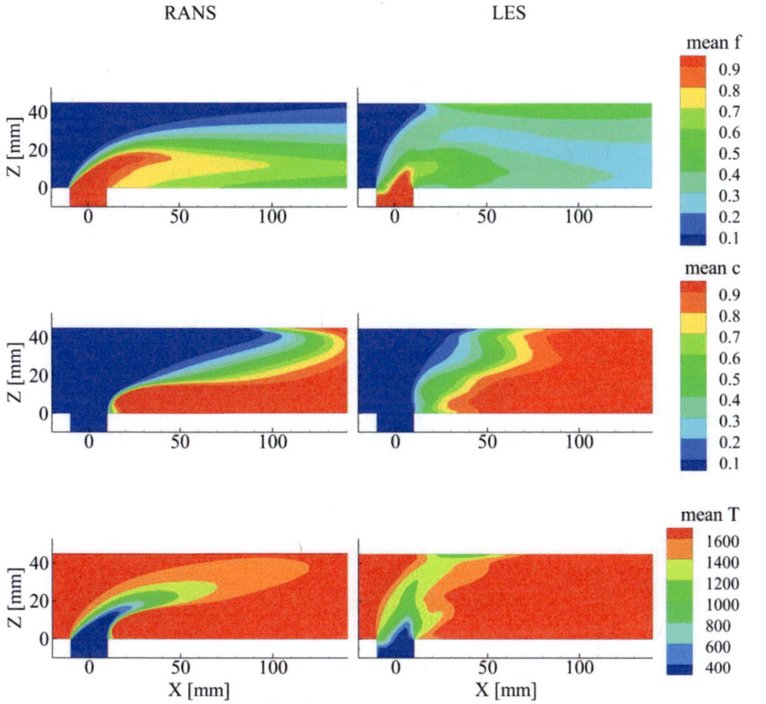

Figure 8.10: Contour of mean mixture fraction f; mean reaction progress variable c and mean temperature T at the symmetry plane, $y = 0$, for RANS and LES simulations with equivalence ratio $\phi = 1.02$.

The comparison of the measurements with the simulations needs to be done with caution. The chemiluminescence emitted by OH* radicals cannot be directly compared to the absolute concentration of OH, as it would require the knowledge of the detailed chemistry of the system involving OH* along with the local temperature. However, the measured OH* chemiluminescence provided qualitative information about the shape of the combustion zone, which is compared to the results computed by the simulations. In the measurement depicted in Figure 8.2 it is shown that at $\phi = 1.02$, the combustion zone stabilizes at some point upstream of the optical window and the reaction front closes at $x = 130$ mm.

To analyze the mean flame front position in the simulations depicted in Figures 8.11 and 8.12, the gradient of the mean reaction progress variable in the axial direction dc/dx is used to represent the flame front. The chemiluminescence measurements already shown in Figure 8.2 are also included in Figure 8.11 and 8.12 to facilitate the comparison with the simulations. It should be noted that the measurements acquired the chemiluminescence over the whole flame volume, while only the cross section of the simulations is being displayed; however, a qualitative comparison is still valid. The measurements show that for the lean mixture (ϕ = 0.55), the flame is stabilized in the region upstream of the optical window and the flame front does not close within the optical window. Both RANS and LES show a similar behavior, with the flame stabilizing slightly downstream of the methane injector and spanning the whole region seen in the picture. For the stoichiometric mixture (ϕ = 1.02) the measurement shows that the flame stabilizes near the methane injector and the flame front closes at x = 120 mm. The simulations show very different behaviors. While the LES agrees qualitatively with the measurements, with the flame front closing at x = 110 mm, the RANS simulation shows a flame front with a radically different shape. These results corroborate with the analysis of Figures 8.9 and 8.10.

8.5 Concluding thoughts

The experiments have shown that a variation in the equivalence ratio significantly affected the location of the secondary stage combustion zone. For lean mixture conditions, the secondary stage combustion zone stabilized well away from the position of the injector, while at stoichiometric conditions it was located closer to the injector.

The flame was embedded in a strong turbulent flow where auto-ignition and quenching are important, which poses a significant challenge for the reaction modeling. The presumed JPDF turbulent reaction model, which has been proven to be a reliable model for these challenging conditions, was successfully coupled with the simulations.

The experimental rig was numerically modeled using steady state RANS and unsteady LES simulations. The predictions of the location of the secondary stage combustion zone by the LES showed good qualitative agreement with the measurements, while the agreement of the RANS simulations was less satisfactory, especially the simulation with stoichiometric mixture (equivalence ratio $\phi = 1.02$).

The causes of the large deviation between the steady state RANS and the LES were studied. The RANS simulations neglect two important sources of unsteadiness, which have a direct impact in the description of the mixing. Besides neglecting the development of large-scale coherent structures, also the acoustic resonance in the methane injector could not be reproduced by the RANS simulations, with severe consequences for the mixing and consequently for the simulation of the whole combustion system. The analysis has shown that time-resolved computational methods like LES are required to model such complex reacting flows.

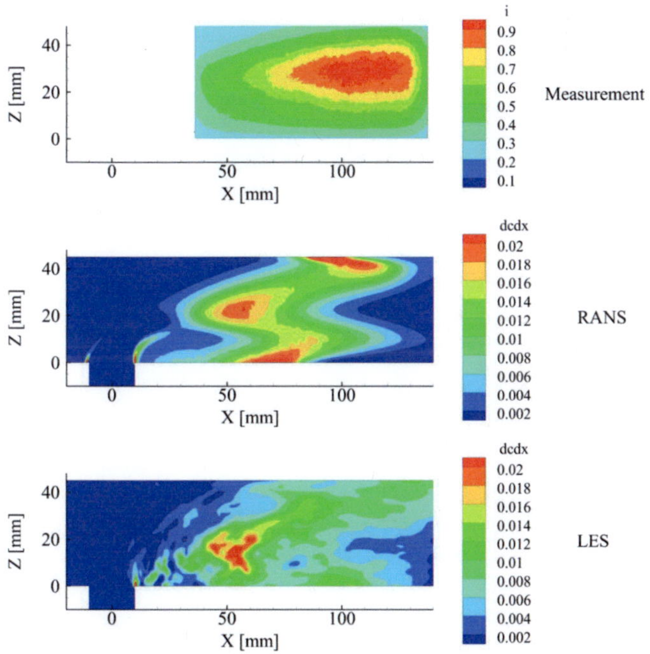

Figure 8.11: Comparison of the measured mean OH* chemiluminescence with the gradient of mean reaction progress variable dc/dx (bottom) of RANS and LES simulations for equivalence ratio $\phi = 0.55$.

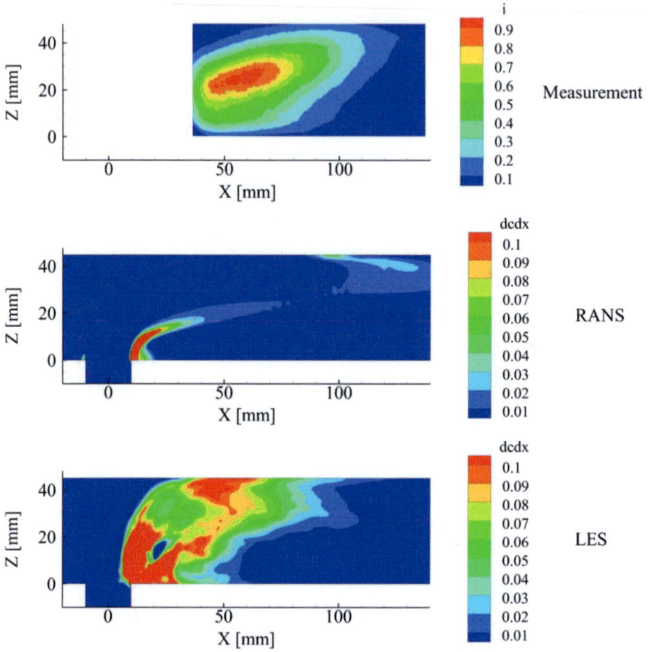

Figure 8.12: Comparison of the measured mean OH* chemiluminescence with the gradient of mean reaction progress variable dc/dx of RANS and LES simulations for equivalence ratio $\phi = 1.02$.

9 Conclusions

The analysis of turbulent mixing in complex turbulent flows is a challenging task. Following the terminology of Eckart [25], turbulent mixing is a three-stage process of entrainment, stirring, and diffusion. While the stirring and the diffusion are usually modeled together by turbulence models, the entrainment remains out of their range. The reason is that in turbulent free shear flows, the entrainment is promoted mainly by the large-scale coherent structures typical for these flows. These coherent structures, defined in Chapter 3, are not turbulent in nature and highly geometry dependent. The derivation of a general model of the effect of coherent structures is a very complex task. Fortunately, unsteady simulations using the Reynolds-averaged Navier Stokes (RANS) or the large eddy simulation (LES) frameworks can resolve the coherent structures directly, accounting for their influence over the time-averaged field.

The intermittency and the conditional averaging techniques proved themselves to be an appropriate framework to account for the influence of large-scale structures on turbulent flows. These tools could quantify successfully quantify the different influence of the coherent structures on the velocity and on the passive scalar fields.

The free jet, a flow configuration extensively studied in the literature, has been used to test and validate the numerical tools and the different simulation methodologies. The LES has successfully predicted the vortex rings originated by the Kelvin-Helmholtz instability, which are the most important coherent structures of the jet flow. The results of the LES agreed well with the major characteristics of the free jet, including the decay of the centerline velocity and of the passive scalar, the jet half-width and the turbulence profiles. The conditional averages of the velocity and the passive scalar have shown the same trends as the measurements of Antonia *et al.* [2] for a similar configuration, with the influence of the intermittency being more pronounced for the passive scalar than for the velocity.

Despite the complexity arising from the coherent structures of the flow, steady state RANS simulations have been able to predict important features of the free jet with good accuracy. The condition for a good simulation is the right choice of turbulence model and constants. The best overall results with RANS simulations have been obtained using the k-ε turbulence model and a turbulent Schmidt number σ_t of 0.7. There is, however, a systematic discrepancy between the RANS results and the measurements in the high intermittent region, which is more evident in the profiles for passive scalar and can be as high as 50% when approaching the jet edge. This discrepancy is related to the fact that steady state RANS simulations do not resolve coherent structures, and consequently could not reproduce the different influence of the coherent structures on the velocity and passive scalar fields.

The jet in crossflow (JIC) is a more complex flow configuration than the free jet, with a series of large-scale coherent structures and recirculation regions. The quality of the agreement of the simulation results with the measurements is strongly coupled with the description of the coherent structures. The LES has successfully predicted the various coherent structures of the flow, including the counter-rotating vortex pair, the horseshoe vortex and the wake vortices. The agreement of the LES results with the measurements is very good for both mean variables and fluctuation values. The unsteady RANS simulation using the SST turbulence model (USST) could resolve only a fraction of the coherent structures. The resulting fluctuation levels are systematically lower than the measurements, even though the agreement of the mean variables is good. The analysis of the results shows that the agreement of the LES results is better than the USST results not only because the turbulence is better reproduced, but also because the LES simulated the coherent structures in much more detail. The steady state RANS simulations do not resolve the coherent structures at all, neglecting this important source of unsteadiness. It follows that the fluctuation levels predicted by the steady state simulations are significantly lower than the measurements. In the steady state simulations, the agreement of the mean velocity is better than the mean passive scalar. This fact could be explained analyzing the conditional averages of the velocity and the passive scalar. The same trend as in the free jet has been found, with a more pronounced influence of the intermittency on the velocity than in the passive scalar. A detailed analysis has shown that, in the flow regions dominated by coherent structures, the turbulent Schmidt number σ_t does not adequately represent the mixing phenomena present there. Different values of σ_t are needed to represent well the mixing in the jet core and along the jet borders, which is consistent with the fact that the intermittency is higher at the borders.

The tools used for the simulation of mixing in turbulent flows with a high level of accuracy have been developed and tested in free jet and jet in crossflow configurations. These tools were adapted to the simulation of a complex combustion system, in which the turbulent mixing plays a major role in the flame geometry and consequently in the whole system. This system is named High Strain Burner, as the combustible mixture is injected with high velocity (and consequently high strain) into the vitiated air stream. The experiments have shown that the fuel amount in the combustible mixture significantly affected the location of the combustion zone. Steady state RANS and unsteady LES simulations of this combustion system were performed. The predictions of the location of the combustion zone by the LES agreed well with the measurements, while the agreement of the RANS simulations was less satisfactory. The causes of the large deviation between the steady state RANS and the LES were analyzed. The RANS simulations neglect two important sources of unsteadiness, which have a direct impact in the description of the mixing. Besides neglecting the development of large-scale coherent structures, also the acoustic resonance in the methane injector could not be reproduced by the RANS simulations, with severe consequences for the mixing and consequently for the simulation of the whole combustion system. The conclusion is that time-resolved computational methods like LES are required to model such complex reacting flows with good accuracy.

The results have shown that steady state RANS simulations provide good quantitative and qualitative agreement with experimental data when the flow is statistically stationary, *i.e.*, when the influence of large-scale coherent structures or unsteadiness of the mean flow are negligible. However, when the flow is not statistically stationary, with pronounced unsteadiness in the mean flow promoted by large-scale coherent structures or other sources, Reynolds-averaged values do not converge to their time-averages. As this unsteadiness is not turbulent in nature, its influence on the mean flow is not modeled by the turbulence models. Hence, to achieve high-fidelity results for all variables, time dependent simulations are mandatory, which increases the computational cost substantially. The preferred method is the LES, which resolved the coherent structures in much more detail than the unsteady RANS. The work has shown the limitations of the steady-state RANS simulations and acknowledged the need of applying unsteady methods for the calculation of the investigated flow configurations. The responsible reasons and phenomena have been identified, analyzed in detail and their impact clearly illustrated.

A Annex

A.1 Computational resources of a DNS

The simulation of a turbulent flow with direct numerical simulation (DNS) requires all turbulent scales (spatial and temporal) to be resolved, which makes the simulation of technically relevant flows out of scope. An example of how small these scales can be follows.

The smallest scale of a turbulent flow is the Kolmogorov length scale (l_K), that can be estimated [114] using the equation

$$\frac{l_0}{l_K} = \mathrm{Re}_{l_0}^{3/4} \tag{A.1}$$

where l_0 is the integral scale, and the Reynolds number calculated using the integral scale Re_{l_0} is defined as

$$\mathrm{Re}_{l_0} = \frac{\rho u_{rms} l_0}{\mu} \tag{A.2}$$

where u_{rms} is the root-mean-square of the velocity fluctuation u'.

An example of the magnitude of the Kolmogorov length scale can be obtained from the pipe flow that originate the jet. The integral scale l_0 is smaller but has the same magnitude as the pipe diameter, assumed to be $0.1D$. The u_{rms} is set to 5% of the bulk jet velocity U_{jet}. Using a typical condition of our study where ρ = 1.185 kg/m^3, U_{jet} = 37.84 m/s, D = 8 mm and $\mu = 1.831 \cdot 10^{-5}$ kg/ms one can obtain the integral scale Reynolds number

$$\mathrm{Re}_{l_0} = \frac{1.185(1.892)(0.0008)}{1.831 \cdot 10^{-5}} = 97.96$$

and the Kolmogorov length scale

$$l_K = l_0 \mathrm{Re}_{l_0}^{-3/4} = 2.5 \cdot 10^{-5}\mathrm{m}.$$

The grid element size has to be smaller than the Kolmogorov length scale, to be able to capture all the physics of the problem. Assuming that an element size of $1 \cdot 10^{-5}$ m is sufficient, the simulation of only 1 cm of pipe flow needs 50 million hexahedral elements, which clearly exceeds the computational infrastructure at disposal, not to mention the simulation of the whole computational domain.

A.2 Kelvin-Helmholtz instability

The Kelvin-Helmholtz instability occurs in flows with shear layers, in which the flow has at least a turning point. The free shear layer formed by flow separation, for example after a nozzle or behind a bluff body, has such one or more turning points. An unstable shear layer can also be formed between two immiscible, stratified fluids with different speeds. The Kelvin-Helmholtz instability describes the spatial and temporal evolution of a perturbation in the shear layer. The instability of the shear layer is solely due to the flow profile and is practically independent of the viscosity, in contrast to the Tollmien-Schlichting instability, which can arise only through the viscosity-induced temporal variation of the flow profile at a wall.

The Kelvin-Helmholtz instability was first described in 1868, proposed by Hermann von Helmholtz [118]. W. Thomson (Lord Kelvin) [111] formulated and solved in 1871 the problem of instability. Analytical solutions of the Kelvin-Helmholtz instability have been compiled, for example, by Batchelor [3] and Chandrasekhar [15]. The figures used in this section have been adapted from [57].

The objective of this section is to introduce the concept of the Kelvin-Helmholtz instability. More details about the derivation of the equations and their analytical solution can be found in references [3], [15] and [57].

A.2.1 Introduction

Two incompressible, viscosity free fluids flowing in parallel planes with different velocities U_1 and U_2 will be considered in the following section. Such a flow arises for example behind a splitter plate (Figure A.1). After the detachment of the boundary layer a free shear layer is formed between the two flows, which is characterized by a large mean velocity gradient dU/dy. The

Kelvin-Helmholtz instability describes the development of this shear layer under an external perturbation. Considering a reference system moving with $U_m = (U_1 + U_2)/2$ and the idealization that the shear layer is infinitely thin (Figure A.2) leads to the thought experiment shown in Figure A.3, in which the shear layer is redirected stationary upward. Due to the faster flow in the upper area, the Bernoulli equation

$$\frac{1}{2}\rho U^2 + p_{stat} = p_{total} = \text{cte.} \tag{A.3}$$

predicts a static underpressure in the upper area, and overpressure in the lower area. In this stationary view, a destabilizing force acts upwards on the shear layer as a whole. In addition, the asymmetry of the flow velocity above and below the shear layer transports it to the right, which leads to a steepening.

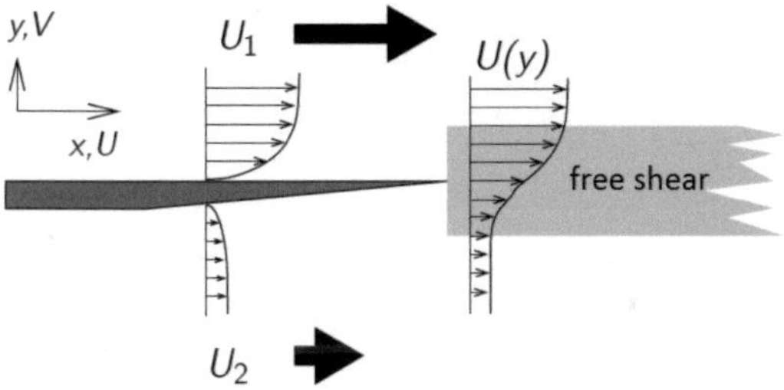

Figure A.1: Development of a free shear flow, adapted from [57].

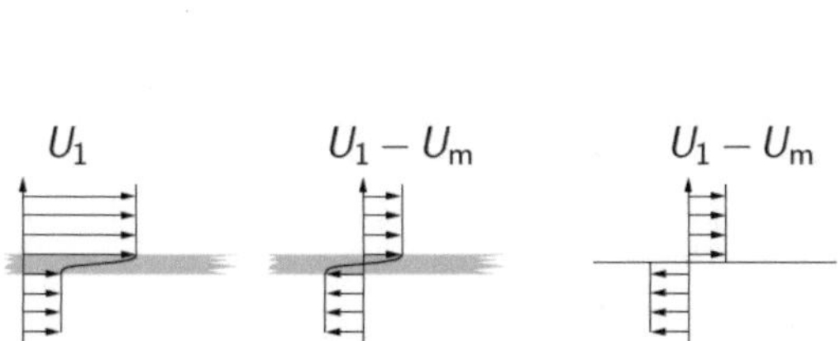

Figure A.2: Coordinate system and idealization of the shear flow, adapted from [57].

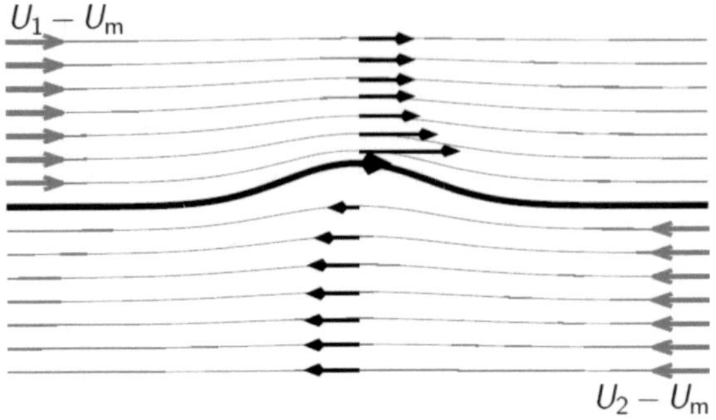

Figure A.3: Potential flow around a perturbation in a shear flow, adapted from [57].

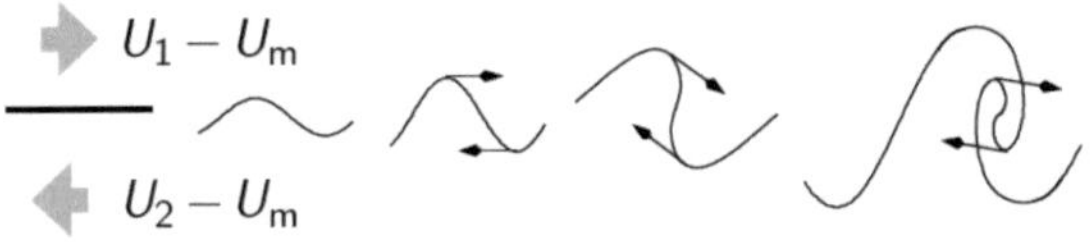

Figure A.4: Roll-up of the shear layer with a finite amplitude, adapted from [57].

If the free shear layer, again in a thought experiment, suffers a sinusoidal deflection (Figure A.4), the amplitude grows in time due to the destabilizing pressure field, and the neighboring strong curved hills and valleys run toward each other, which results in the roll-up of the shear layer. The effect of pressure drop across the shear layer, which leads to its acceleration, has been neglected in this consideration using the steady Bernoulli equation. The transient problem can be completely described by vortex dynamics [3].

The propagation velocity of an infinitesimal, spatially and temporally perturbation is $U_m = (U_1 + U_2)/2$ because of symmetry. In a reference frame in which U_1 is equal to $-U_2$, the perturbation just grows over time and remain at the same location, characterizing an absolute instability. In a reference frame that moves with U_2, the instability flows away from the point of perturbation and is amplified spatially; after a certain time, the instability disappears from the point of perturbation. This is called a convective instability (Figure A.5).

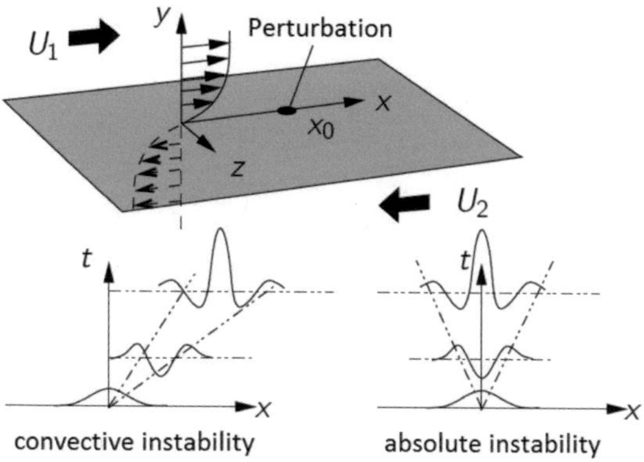

Figure A.5: Spatial and temporal development of a perturbation, adapted from [57].

The concept of the Kelvin-Helmholtz instability can be broadened if one considers various fluids with different densities and surface tensions and involve also the effects of gravity. In principle, this changes only the relationship of the lateral velocity and pressure at the interface between the two fluids. The surface tension and the weight have a stabilizing effect on the Kelvin-Helmholtz instability.

A.2.2 Characteristics and functions

If a laminar flow with the main flow velocity $U(y)$ suffers a sinusoidal perturbation with frequency ω_s and this perturbation is infinitesimal, in terms of classical stability theory, each flow variable ϕ can be represented as a sum of a time-averaged part $\overline{\phi}$ and a perturbing wave with the wavelength $2\pi/\Re\{k\}$ and an excitation $\Im\{k\}$:

$$\phi(x,y,t) = \overline{\phi}(y) + \tilde{\phi} \cdot e^{i(kx-\omega_s t)}. \tag{A.4}$$

From the momentum equation for incompressible flows the amplitude of the transverse velocity can be derived linearizing the Rayleigh equation:

$$\frac{d^2\tilde{v}(y)}{dy^2} = \left(k^2 + \frac{\frac{d^2\overline{u}(y)}{dy^2}}{\overline{u}(y) - \frac{\omega_s}{k}} \right) \cdot \tilde{v}(y), \tag{A.5}$$

where $\tilde{v}(y)$ should vanish for $y \to \pm\infty$. The eigenvalues of this equation result in complex relations between the introduced perturbation frequency ω_s and the complex wave number k. It is also possible to find the time amplification and the oscillation period ω for a given spatial wave number k_s, which will not be considered here.

In Figure A.6, the spatial amplification and wavelength (complex wave number k) as a function of the real interference frequency ω_s is shown. As dimensionless excitation frequency the Strouhal number is used

$$\mathrm{Sr} = \frac{\omega_s \delta / 2}{|U_1 - U_2|}, \tag{A.6}$$

where δ is the characteristic thickness of the shear layer. For shear layers of finite thickness, there is a window in which the amplification ($-\Im\{k\} > 0$) takes place.

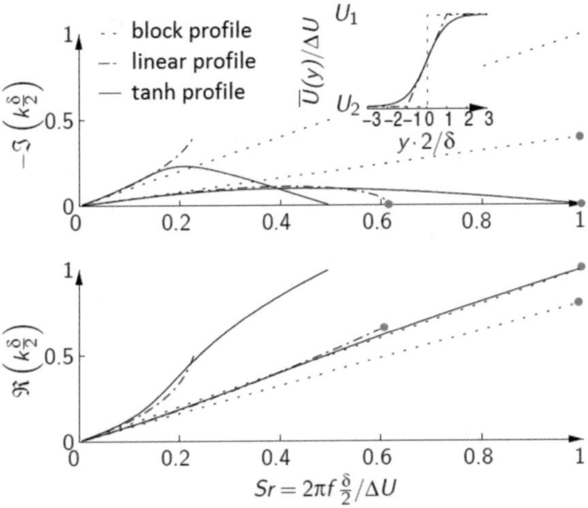

Figure A.6: Relation of dispersion of different flow configurations. The lines with point at their ends are calculated for $U_1 = 3/2U_m$ and $U_2 = 1/2U_m$, the lines without points are valid for $U_1 = 2U_m$ and $U_2 = 0$, adapted from [57].

The flow profile of a free jet leads to two dispersion relations (modes) as shown in Figure A.7.

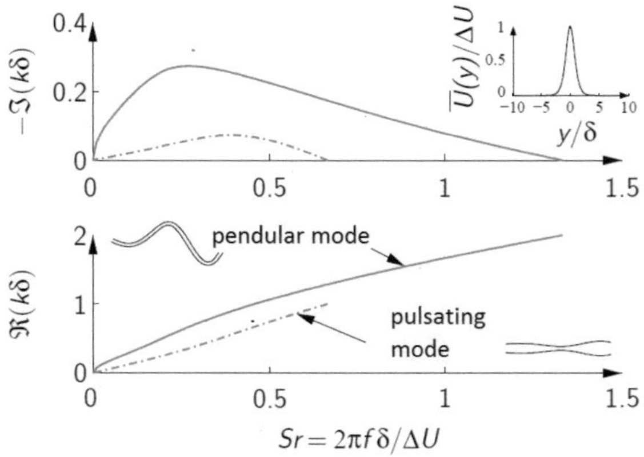

Figure A.7: Pendular and pulsating modes for a free jet with velocity profile $\Delta U/\cosh(y/\partial)^2$, adapted from [57].

A.2.3 Applications

The Kelvin-Helmholtz instability can be observed in the sky when temperature inversions occur, resulting in the typical Kelvin-Helmholtz clouds (Figure A.8). The so-called CAT (clear air turbulence), which is also formed by the Kelvin-Helmholtz instability, can be very dangerous for aircrafts due to the strong vertical velocities generated.

The excitation of water waves by wind, can also be described in part by the Kelvin-Helmholtz instability.

If the splitter plate configuration in Figure A.1 is perturbed by an oscillating pressure field (here by means of two membranes, which move in phase), Kelvin-Helmholtz waves are formed, as the ones shown in Figure A.9, made visible using ink. The same can be seen for a free jet in Figure A.10 (a) and (b).

The flow profile of the wake behind a cylinder in cross flow is absolutely unstable. This results in the Kármán vortex street.

Figure A.8: Kelvin-Helmholtz clouds, adapted from [57].

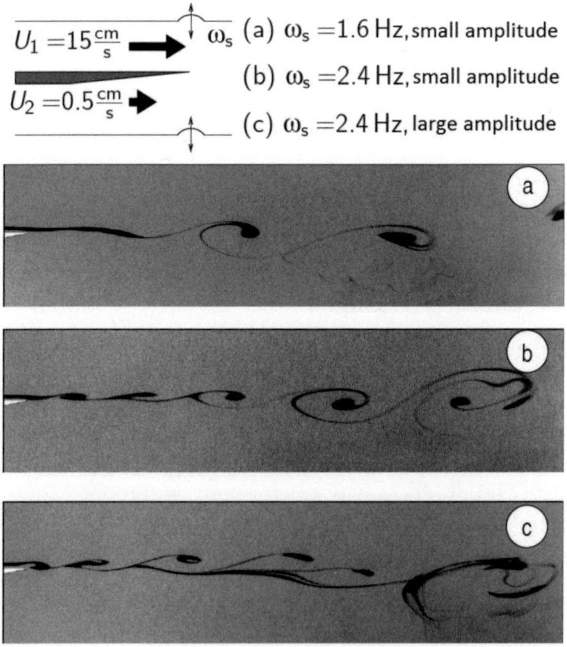

Figure A.9: Artificial excitation of the Kelvin-Helmholtz instability in water at the edge of a plate, adapted from [57].

The Kelvin-Helmholtz instability generates sound when a Kelvin-Helmholtz wave encounters an obstacle. In Figure A.10 (c) a self-excited free jet is seen, with the flow oscillations made visible with ink. The tone produced in this way is used in organ pipes and flutes, in which an oscillator is coupled; the resonance frequency of the oscillator determines the oscillation frequency of the Kelvin-Helmholtz instability. The noise generation is sometimes undesirable, for example the rumble of an open train or car window.

When injecting a mixture through a nozzle into a space filled with a medium that can mix well with this jet mixture, the energy of the turbulent mixing can be increased by an artificial excitation of the Kelvin-Helmholtz instability, which can speed up the mixing.

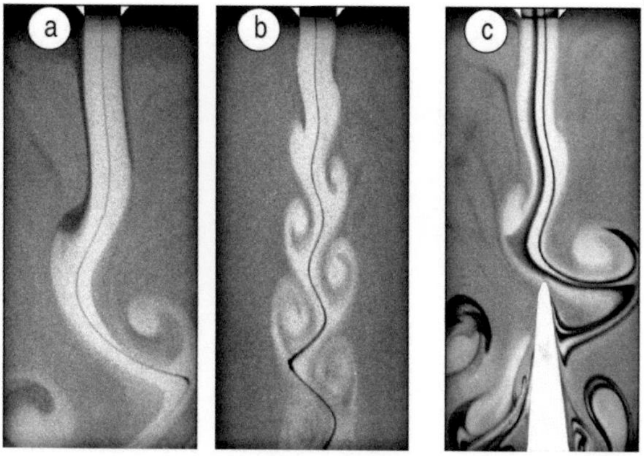

Figure A.10: Convective instability of a free jet with $\Delta U = 5$cm/s: (a) and (b) with artificial excitation $\omega_s = 1$Hz (a) and 2Hz (b), self-excited oscillation of the jet-edge system, adapted from [57].

A.3 Computational resources used for the simulations

Two systems were available to conduct the simulations: the in-house Linux cluster of the Engler-Bunte-Institute, Division of Combustion Technology (EBI-VBT), and the HP XC4000 high performance computer [115] of the federal state Baden-Württemberg, available through the Steinbuch Centre for Computing of the Karlsruher Institute of Technology (KIT).

The parallel performance of the computational systems was assessed by LES simulations using coarse and fine grids. The coarse grid was relatively small, and fitted well into the in-house cluster, while the fine grid was better suited for the HP XC4000 system. The parallel performance of both systems was very good [32].

The in-house cluster consists of various subsystems:

- 1 12-core node with two Intel Xeon X7460 processors at 2.663 GHz.
- 19 4-core nodes with Intel Core2 Q9550 processors at 2.83 GHz.
- 22 4-core nodes with Intel i7 860 processors at 2.8 GHz.

The 12-core node was used to generate the grids, mainly because of the available 32 GB of memory. The simulations have been conducted in the 4-core Intel i7 860 nodes.

Table A.1 shows a comparison of the main aspects of the two systems.

	in-house cluster	HP XC4000
Processors	Intel i7 860, 2.8 GHz	AMD Opteron, 2.6 GHz
Cores per node	1 x 4 cores	2 x 2 cores
Memory per node	8GB	16GB
Interconnect	Gigabit Ethernet	InfiniBand 4X
MPI version	OpenMPI	HPMPI

Table A.1: Comparison of the cluster systems.

The main bottleneck of the in-house cluster is the Gigabit Ethernet interconnect, which limits the scalability of the simulations. The scalability is above 90% until 4 nodes, being significantly degraded when using more nodes. In contrast, the HP XC4000 system showed good scalability, with over 90% efficiency, up to 64 nodes (256 processors), the maximum number of nodes tested.

Table A.2 shows an overview of the computational time for different simulations of the jet in crossflow using the in-house cluster. The performance of the HP XC4000 system is in general 25% worse per core, which is a consequence of the older AMD Opteron processors. The steady state RANS simulation was simulated until convergence; the unsteady simulations have been simulated for a total time of 0.15 seconds. The difference in computing time is huge. The steady state RANS converges in 1.5 days per core; using 8 cores (2 nodes) the time needed decreases to slightly more than 4 hours. The URANS and the LES have almost the same performance using the same grid, demanding slightly more than one year computational time on a single core. The fine grid used for the last LES is 5 times larger than the coarse one, and also uses a time step that is one half of the time step used with the coarser grid, demanding approximately ten times more computational time than the LES using the coarse grid. One processor would take more than ten years to compute this simulation. Fortunately, using 256 parallel nodes in the HP XC4000 system reduced the computing time to around two weeks.

Simulation	Grid size (elements)	computational time (days/core)
Steady state RANS	1.5 million	1.5
Unsteady RANS	1.5 million	394
LES	1.5 million	377
LES	7.5 million	3 712

Table A.2: Overview of the computational time for different simulations of the jet in crossflow.

It should be noted that there is room for reduction of the computational time of the unsteady RANS simulations. The current simulations have employed the PISO time discretization technique, which limits the Courant number of the simulation to values smaller than unity and consequently limits the time step size. As the LES formalism already restricts the acceptable Courant number, this restriction is not decisive. The unsteady RANS, on the other hand, could use larger time steps without compromising the accuracy when using alternative time discretization techniques.

Bibliography

[1] P. Anderson, J. C. LaRue, and P. A. Libby. Preferential entrainment in a 2-dimensional turbulent jet in a moving stream. *Physics of Fluids*, 22(10):1857–1861, 1979.

[2] R. A. Antonia, A. Prabhu, and S. E. Stephenson. Conditionally sampled measurements in a heated turbulent jet. *Journal of Fluid Mechanics*, 72(3):455–480, 1975.

[3] G.K. Batchelor. *An Introduction To Fluid Dynamics*. Cambridge University Press, 1967.

[4] R. B. Bird, W. Stewart, and E. Lightfoot. *Transport Phenomena*. John Wiley and Sons, 2002.

[5] R. Borghi. Turbulent combustion modelling. *Progress in Energy and Combustion Science*, 14:245–292, 1988.

[6] J. P. Boris and D. L. Book. Flux-corrected transport. I. SHASTA, a fluid transport algorithm that works. *J. Comp. Physics*, 11:38–69, 1973.

[7] P. Bradshaw. *Turbulence, Topics in Applied Physics, Vol.12*. Springer Verlag, 1978.

[8] N. Branley and W. P. Jones. Large eddy simulation of a turbulent non-premixed flame. *Combustion and Flame*, 127:1914–1934, 2001.

[9] J. E. Broadwell and R. E. Breidenthal. Structure and mixing of a transverse jet in incompressible flow. *Journal of Fluid Mechanics*, 148:405–412, 1984.

[10] J. E. Broadwell and M. G. Mungal. Large-scale structures and molecular mixing. *Physics of Fluids A*, 3(5):1193–1206, 1991.

[11] G. L. Brown and A. Roshko. Density effects and large structure in turbulent mixing layers. *Journal of Fluid Mechanics*, 64:775–816, 1974.

[12] O. Brunn, F. Wetzel, P. Habisreuther, and N. Zarzalis. Investigation of a combustor using a presumed JPDF reaction model applying radiative heat loss by the Monte Carlo method. In *Proceedings of the 25th Congress of the International Council of the Aeronautical Sciences (ICAS)*, 2006.

[13] R. Camussi, G. Guj, and A. Stella. Experimental study of a jet in a crossflow at very low Reynolds number. *Journal of Fluid Mechanics*, 454:113–144, 2002.

[14] C. Cárdenas, R. Suntz, J.A. Denev, and H. Bockhorn. Two-dimensional estimation of Reynolds-fluxes and -stresses in a Jet-in-Crossflow arrangement by simultaneous 2D-LIF and PIV. *Applied Physics B: Lasers and Optics*, 88(4):581–591, 2007.

[15] S. Chandrasekhar. *Hydrodynamic and hydromagnetic stability*. Oxford University Press, 1961.

[16] S. L. V. Coelho and J. C. R. Hunt. The dynamics of the near field of strong jets in crossflows. *Journal of Fluid Mechanics*, 200:95–120, 1989.

[17] S. Corrsin. Investigation of flow in an axially symmetrical heated jet of air. *NACA, Wash. Wartime Report*, W-94, 1943.

[18] C. Cárdenas, J. Sedlmaier, N. Zarzalis, R. J. Valdes, and W. Krebs. Measurement of a benchmarking jet in crossflow configuration under highly turbulent conditions. In *Proceedings of ASME Turbo Expo 2011: Power for Land, Sea and Air*. ASME, June 06-10 2011. Paper number GT2011-45262.

[19] W. J. A. Dahm and P. E. Dimotakis. Measurements of entrainment and mixing in turbulent jets. *AIAA Journal*, 25(9):1216–1223, 1987.

[20] E. de Villiers. *The Potential of Large Eddy Simulation for the Modeling of Wall Bounded Flows*. PhD thesis, Imperial College of Science, Technology and Medicine, 2006.

[21] M. Dianat, Z. Yang, D. Jiang, and J. J. McGuirk. Large eddy simulation of scalar mixing in a coaxial confined jet. *Flow Turbulence and Combustion*, 77:205–227, 2006.

[22] P. E. Dimotakis. Turbulent mixing. *Annual Review of Fluid Mechanics*, 37:329–356, 2005.

[23] P. Domingo, L. Vervischa, and D. Veynante. Large-eddy simulation of a lifted methane jet flame in a vitiated coflow. *Combustion and Flame*, 152:415–432, 2008.

[24] D. R. Dowling and P. E. Dimotakis. Similarity of the concentration field of gas-phase turbulent jets. *Journal of Fluid Mechanics*, 218:109–141, 1990.

[25] C. Eckart. An analysis of the stirring and mixing processes in incompressible fluids. *Journal of Marine Research*, 7(3):265–275, 1948.

[26] A. Favre. Statistical equations of turbulent gases. In *Problems of hydrodynamics and continuum mechanics*, pages 231–266. SIAM, Philadelphia, 1969.

[27] J. Fröhlich. *Large Eddy Simulation turbulenter Strömungen*. Teubner, 2006.

[28] T. F. Fric and A. Roshko. Vortical structure in the wake of a transverse jet. *Journal of Fluid Mechanics*, 279:1–47, 1994.

[29] E. Gabriel, G. E. Fagg, G. Bosilca, T. Angskun, J. J. Dongarra, J. M. Squyres, V. Sahay, P. Kambadur, B. Barrett, A. Lumsdaine, R. H. Castain, D. J. Daniel, R. L. Graham, and T. S. Woodall. Open MPI: Goals, concept, and design of a next generation MPI implementation. In *Proceedings of the 11th European PVM/MPI Users' Group Meeting*, pages 97–104, Budapest, Hungary, September 2004.

[30] F. C. C. Galeazzo, G. Donnert, P. Habisreuther, N. Zarzalis, R. J. Valdes, and W. Krebs. Measurement and simulation of turbulent mixing in a jet in crossflow. In *Proceedings of ASME Turbo Expo 2010: Power for Land, Sea and Air*, pages Paper number GT2010–22709. ASME, June 14-18 2010.

[31] F. C. C. Galeazzo, G. Donnert, P. Habisreuther, N. Zarzalis, R. J. Valdes, and W. Krebs. Measurement and simulation of turbulent mixing in a jet in crossflow. *Journal of Engineering for Gas Turbines and Power*, 133(3):061504.1–10, 2011.

[32] F. C. C. Galeazzo, P. Habisreuther, and N. Zarzalis. Large eddy simulations of a jet in crossflow. In *High Performance Computing in Science and Engineering '10*, pages 327–337. Springer Berlin Heidelberg, 2011.

[33] F. C. C. Galeazzo, M. Kern, P. Habisreuther, N. Zarzalis, and C. Beck. Simulation of a lifted flame in a vitiated air environment. In *Proceedings of the European Combustion Meeting 2011, British Section of the Combustion Institute, Cardiff, UK*, 2011.

[34] F. C. C. Galeazzo, C. Prathap, M. Kern, P. Habisreuther, N. Zarzalis, and C. Beck. Investigations of an axially staged combustion system at elevated pressure conditions. In *25. Deutscher Flammentag, Karlsruhe, Germany*, pages 265–270, 2011.

[35] F. C. C. Galeazzo, C. Prathap, M. Kern, P. Habisreuther, N. Zarzalis, C. Beck, W. Krebs, and B. Wegner. Investigation of a flame anchored in crossflow stream of vitiated air at elevated pressures. In *Proceedings of ASME Turbo Expo 2012: Power for Land, Sea and Air*. ASME, June 11-15 2012. Paper number GT2012-69632.

[36] M. Germano, U. Piomelli, P. Moin, and W. Cabot. A dynamic subgrid-scale eddy viscosity model. *Physics of Fluids A*, 3:1760 – 1765, 1991.

[37] D. G. Goodwin. An open-source, extensible software suite for CVD process simulation. In *Chemical Vapor Deposition XVI and EUROCVD 14, ECS Proceedings, M. Allendorf, F. Maury, and F. Teyssandier, editors*, volume 8, pages 155–162. The Electrochemical Society, 2003.

[38] D. Großschmidt, A. Hoffmann, P. Habisreuther, H. Bockhorn, and S. Hohmann. Modeling turbulence/chemistry interactions for industrial application using assumed PDF methods. In *Proceedings of the European Combustion Meeting (ECM 2003)*, 2003.

[39] J. O. Hinze. *Turbulence*. MC Graw HILL, 1975.

[40] J. O. Hirschfelder, C. F. Curtiss, and R. B. Byrd. *Molecular theory of gases and liquids*. John Wiley and Sons, New York, 1969.

[41] A. Hoffmann, D. Großschmidt, M. Zajadatz, W. Leuckel, and H. Bockhorn. Validierung eines JPDF-Modells sowie einer semi-globalen Reaktionskinetik anhand detaillierter Feldmessungen hochturbulenter vorgemischter Freistrahlflammen. In *VDI-GET-Fachtagung 'Verbrennung und Feuerungen-20. Deutscher Flammentag'*, 2001. VDI-Bericht Nr. 1629.

[42] A. K. M. F. Hussain. Coherent structures - reality and myth. *Physics of Fluids*, 26(10):2816–2850, 1983.

[43] A. K. M. F. Hussain. Coherent structures and turbulence. *Journal of Fluid Mechanics*, 173:303–356, 1986.

[44] J. H. Hussein, S. P. Capp, and W. K. George. Velocity measurements in a high-Reynolds-number, momentum-conserving, axisymmetric, turbulent jet. *Journal of Fluid Mechanics*, 258:31–75, 1994.

[45] R. I. Issa. Solution of the implicitly discretised fluid flow equations by operator-splitting. *Journal of Computational Physics*, 62:40–65, 1986.

[46] H. Jasak. *Error Analysis and Estimation for the Finite Volume Method with Applications to Fluid Flows.* PhD thesis, Imperial College of Science, Technology and Medicine, 1996.

[47] W. P. Jones and B. E. Launder. The prediction of laminarization with a two-equation model of turbulence. *International Journal of Heat and Mass Transfer*, 15:301–314, 1972.

[48] A. R. Karagozian. An analytical model for the vorticity associated with a transverse jet. *AIAA Journal*, 24:429–436, 1986.

[49] G. Karypis and V. Kumar. A fast and high quality multilevel scheme for partitioning irregular graphs. *SIAM Journal on Scientific Computing*, 20(1):359–392, 1999.

[50] R. M. Kelso, T. T. Lim, and A. E. Perry. An experimental study of round jets in cross-flow. *Journal of Fluid Mechanics*, 306:111–144, 1996.

[51] M. A. Kempf. LES validation from experiments. *Flow, Turbulence and Combustion*, 80:351–373, 2008.

[52] M. Kern, P. Fokaides, P. Habisreuther, and N. Zarzalis. Applicability of a flamelet and presumed JPDF 2-domain-1-step-kinetic turbulent reaction model for the simulation of a lifted swirl flame. In *Proceedings of ASME Turbo Expo 2009*. ASME, 2009. Paper number GT2009-5943.

[53] M. Klein, A. Sadiki, and J. Janicka. A digital filter based generation of inflow data for spatially developing direct numerical or large eddy simulations. *Journal of Computational Physics*, 186:652–665, 2003.

[54] A. N. Kolmogorov. The local structure of turbulence in incompressible viscous fluid for very large Reynolds numbers. *Dokl. Akad. Nauk SSSR*, 30:299–303, 1941.

[55] F. P. Kärrholm. *Numerical Modelling of Diesel Spray Injection, Turbulence Interaction and Combustion.* PhD thesis, Department of Applied Mechanics, Chalmers University of Technology, Göteborg, Sweden, 2008.

[56] K. K. Kuo. *Principles of Combustion, Second Edition.* John Wiley and Sons, Inc., 2005.

[57] T. Kurz. Praktikum für Fortgeschrittene - Kelvin-Helmholtz-Instabilität. Technical report, Drittes Physikalisches Institut der Fakultät für Physik an der Georg-August-Universität Göttingen, 2003. www.dpi.physik.uni-goettingen.de/Praktika/Fortgeschrittenen/Versuche/V254/kelvin.pdf.

[58] M. Lauer and T. Sattelmayer. On the adequacy of chemiluminescence as a measure for heat release in turbulent flames with mixture gradient. *Journal of Engineering for Gas Turbines and Power*, 132(6):061502, 2010.

[59] B. E. Launder and D. B. Spalding. The numerical computation of turbulent flows. *Comp. Meth. Appl. Mech. Eng.*, 3:269–289, 1974.

[60] C. J. Lawn. Lifted flames on fuel jets in co-flowing air. *Progress in Energy and Combustion Science*, 35:1–30, 2009.

[61] A. Leonard. Energy cascade in large-eddy simulations of turbulent fluid flows. *Advances in Geophysics*, 18(1):237 – 248, 1975.

[62] B. P. Leonard. A stable and accurate convective modelling procedure based on quadratic upstream interpolation. *Comp. Meth. Appl. Mech. Engineering*, 19:59–98, 1979.

[63] P. A. Libby. *Introduction to Turbulence.* Taylor & Francis, 1996.

[64] D. K. Lilly. A proposed modification of the germano subgrid-scale closure method. *Phys. Fluids A*, 4:633–635, 1992.

[65] T. T. Lim. Flow gallery. http://serve.me.nus.edu.sg/limtt/, 2011.

[66] T. T. Lim, T. H. New, and S. C. Luo. On the development of large-scale structures of a jet normal to a cross flow. *Physics of Fluids*, 13:770–775, 2001.

[67] T. Lischer, G. Donnert, F. C. C. Galeazzo, P. Habisreuther, N. Zarzalis, R. Valdes, and W. Krebs. Simultaneous velocity and concentration measurements using laser-optical measurement methods in comparison with Reynolds Averaged Navier-Stokes models. In *Proceedings of the 12th International Symposium on Transport Phenomena and Dynamics of Rotating Machinery*. ISROMAC, February 2008. Paper number ISROMAC12-2008-20112.

[68] E. J. List. Turbulent jets and plumes. *Annual Review of Fluid Mechanics*, 14:18–212, 1982.

[69] F. C. Lockwood and A. S. Naguib. The prediction of the fluctuations in the properties of free, round-jet, turbulent, diffusion flames. *Combustion and Flame*, 24:109–124, 1975.

[70] T. Lund, X. Wu, and D. Squires. Generation of turbulence inflow data for spacially-developing boundary layer simulations. *J. Comp. Physics*, 140:233–258, 1998.

[71] K. M. Lyons. Toward an understanding of the stabilization mechanisms of lifted turbulent jet flames: Experiments. *Progress in Energy and Combustion Science*, 33:211–231, 2007.

[72] R. J. Margason. Fifty years of jet in cross flow research. In *AGARD, Computational and Experimental Assessment of Jets in Cross Flow*, chapter 1. AGARD CP-534, Winchester, UK, 1993.

[73] F. R. Menter. Two-equation eddy-viscosity models for engineering applications. *American Institute of Aeronautics and Astronautics Journal*, 32:1598–1605, 1994.

[74] F. R. Menter and Y. Egorov. A scale-adaptive simulation model using two-equation models. Technical Report AIAA 2005-1095, American Institute of Aeronautics and Astronautics, 2005.

[75] D. Mi, D. S. Nobes, and G. J. Nathan. Influence of exit conditions on the passive scalar field of an axisymmetric free jet. *Journal of Fluid Mechanics*, 432:91–125, 2001.

[76] J. Michel, O. Colin, C. Angelberger, and D. Veynante. Using the tabulated diffusion flamelet model ADF-PCM to simulate a lifted methane-air jet flame. *Combustion and Flame*, 156(7):1318–1331, 2009.

[77] B. R. Morton and A. Ibbetson. Jets deflected in a crossflow. *Experimental Thermal and Fluid Science*, 12(2):112 – 133, 1996.

[78] M. G. Mungal and D. K. Hollingsworth. Organized motion in a very high Reynolds number jet. *Physics of Fluids A*, 1(10):1615–1623, 1989.

[79] S. Muppidi and K. Mahesh. Study of trajectories of jets in crossflow using direct numerical simulations. *Journal of Fluid Mechanics*, 530:81–100, 2005.

[80] S. Muppidi and K. Mahesh. Two-dimensional model problem to explain counter-rotating vortex pair formation in a transverse jet. *Physics of Fluids*, 18:085103.1–9, 2006.

[81] T. H. New, T. T. Lim, and S. C. Luo. Elliptic jets in cross-flow. *Journal of Fluid Mechanics*, 494:119–140, 2003.

[82] C. Olbricht, F. Hahn, A. Ketelheun, and J. Janicka. Strategies for presumed PDF modeling for LES with premixed flamelet-generated manifolds. *Journal of Turbulence*, 11(38):1–18, 2010.

[83] G. Ooms. A new method for the calculation of the plume of gases emitted by a stack. *Atmospheric environment*, 6(12):899–909, 1967.

[84] OpenCFD Ltd. *OpenFOAM User Guide, Version 1.5*, 2008.

[85] OpenCFD Ltd. *OpenFOAM User Guide, Version 2.0.0*, 2011.

[86] N. R. Panchapakesan and J. L. Lumley. Turbulence measurements in axisymmetric jets of air and helium. Part 1. Air jet. *Journal of Fluid Mechanics*, 246:197–223, 1993.

[87] S. V. Patankar. *Numerical heat transfer and fluid flow*. Hemisphere Publishing Corporation, 1981.

[88] S. V. Patankar and D. B. Spalding. A calculation procedure for heat, mass and momentum transfer in three-dimensional parabolic flows. *International Journal of Heat and Mass Transfer*, 15:1787–1806, 1972.

[89] N. Peters. *Turbulent Combustion*. Cambridge University Press, 2000.

[90] W. M. Pitts. Assessment of theories for the behavior and blowout of lifted turbulent jet diffusion flames. In *Twenty-Second Symposium (International) on Combustion*, pages 809–816, 1988.

[91] T. Poinsot and D. Veynante. *Theoretical and Numerical Combustion, Second Edition*. R.T. Edwards, Inc., 2005.

[92] T. J. Poinsot and S. K. Lele. Boundary conditions for direct simulations of compressible viscous flows. *Journal of Computational Physics*, 101:104–129, 1992.

[93] S. B. Pope. *Turbulent Flows*. Cambridge University Press, 2000.

[94] L. Prandtl. Über ein neues Formelsystem für die ausgebildete Turbulenz. *Nachrichten der Akademie der Wissenschaften, Mathematisch Physikalische Klasse*, 6:6–19, 1945.

[95] C. Prathap, F. C. C. Galeazzo, P. Kasabov, P. Habisreuther, N. Zarzalis, C. Beck, W. Krebs, and B. Wegner. Analysis of NOx formation in an axially staged combustion system at elevated pressure conditions. In *Proceedings of ASME Turbo Expo 2011*, pages Paper GT2011–45239, 2011.

[96] C. Prathap, F. C. C. Galeazzo, P. Kasabov, P. Habisreuther, N. Zarzalis, C. Beck, W. Krebs, and B. Wegner. Analysis of NOx formation in an axially staged combustion system at elevated pressure conditions. *Journal of Engineering for Gas Turbines and Power*, 134(3):031507.1–8, 2012.

[97] G. D. Raithby. A critical evaluation of upstream differencing applied to problems involving fluid flow. *Comp. Meth. Appl. Mech. Engineering*, 9:75–103, 1976.

[98] G. D. Raithby. Skew-upstream differencing schemes for problems involving fluid flow. *Comp. Meth. Appl. Mech. Engineering*, 9:153–164, 1976.

[99] O. Reynolds. On the dynamical theory of incompressible viscous fluids and the determination of the criterion. *Philosophical Transactions of the Royal Society of London. A*, 186:123–164, 1895.

[100] A. Rivero, J. A. Ferre, and F. Giralt. Organized motions in a jet in crossflow. *Journal of Fluid Mechanics*, 444:117–149, 2001.

[101] W. Rodi. A review of experimental data of uniform density free turbulent boundary layers. In *Studies in Convection, ed. B. E. Launder*, pages 79–165. Academic, 1975.

[102] M. M. Rogers and R. D. Moser. Direct simulation of a self-similar turbulent mixing layer. *Phys. Fluids*, 6:903–923, 1994.

[103] J.U. Schlüter and T. Schönfeld. LES of jets in crossflow and its application to a gas turbine burner. *Flow, Turbulence and Combustion*, 65(2):177–203, 2000.

[104] J. Smagorinsky. General circulation experiments with the primitive equations: I. the basic equations. *Mon. Weather Rev.*, 91:99–164, 1963.

[105] G. P. Smith, D. M. Golden, M. Frenklach, N. W. Moriarty, B. Eiteneer, M. Goldenberg, C. T. Bowman, R. K. Hanson, S. Song, W. C. Gardiner Jr., V. V. Lissianski, and Qin Z. GRI-mech 3.0. http://www.me.berkeley.edu/gri_mech/.

[106] S. H. Smith and M. G. Mungal. Mixing, structure and scaling of the jet in crossflow. *Journal of Fluid Mechanics*, 357:83–122, 1998.

[107] P. R. Spalart. Direct simulation of a turbulent boundary layer up to $R\theta = 1410$. *Journal of Fluid Mechanics*, 187:61–98, 1998.

[108] D. B. Spalding. A single formula for the law of the wall. *Transactions of the American Society of Mechanical Engineers, Series E*, 28:455 – 458, 1961.

[109] S. Tavoularis and S. Corrsin. Experiments in nearly homogenous turbulent shear-flow with a uniform mean temperature-gradient. part 1. *Journal of Fluid Mechanics*, 104:311–347, 1981.

[110] H. Tennekes and J. L. Lumley. *A First Course in Turbulence*. MIT Press, 1972.

[111] W. Thomson (Lord Kelvin). Hydrokinetic solutions and observations. *Philosophical Magazine*, 42(4):362–377, 1871.

[112] A. A. Townsend. Local isotropy in the turbulent wake of a cylinder. *Australian Journal Of Scientific Research Series A-Physical Sciences*, 1:161–174, 1948.

[113] A. Travin, M. Shur, M. Strelets, and P. Spalart. Detached-eddy simulations past a circular cylinder. *Flow, Turbulence and Combustion*, 63:293–313, 2000.

[114] S. R. Turns. *An Introduction to Combustion*. McGraw-Hill, 2000.

[115] Universität Karlsruhe (TH), Rechenzentrum. *HP XC4000 User Guide*, 2009.

[116] E. Van den Bulck. Research group thermotech, K. U. Leuven. www.mech.kuleuven.be/en/tme/research/Thermotech/FlowCombust/, 2011.

[117] J. A. van Oijen, F. A. Lammers, and L. P. H. de Goey. Modeling of premixed laminar flames using flamelet generated manifolds. *Combustion Science and Technology*, 127:2124–2134, 2001.

[118] H. von Helmholtz. Flüssigkeitsbewegungen. *Monatsberichte Königlich Preussische Akad. Wiss. Berlin*, 23:215–28, 1868.

[119] A. W. Vreman, B. A. Albrecht, J. A. van Oijen, L. P. H. de Goey, and R. J. M. Bastiaans. Premixed and Non-premixed generated manifolds in large-eddy simulation of Sandia flame D and F. *Combustion and Flame*, 153:394–415, 2008.

[120] F. Wetzel, P. Habisreuther, and N. Zarzalis. Numerical investigation of lean blow out of a model gas turbine combustion chamber using a presumed JPDF-reaction model by taking heat loss processes into account. In *Proceedings of ASME Turbo Expo 2006*. ASME, 2006. Paper number GT2006-90064.

[121] D. C. Wilcox. *Turbulence Modeling for CFD, Third Edition*. DCW Industries, Inc., 2006.

[122] F. A. Williams. *Combustion Theory*. Addison Wesley Publishing, 1988.

[123] I. Yimer, I. Campbell, and L-Y. Jiang. Estimation of the turbulent Schmidt number from experimental profiles of axial velocity and concentration for high-Reynolds-number jet flows. *Canadian Aeronautics and Space Journal*, 48(3), 2002.

[124] D. You and P. Moin. A dynamic global-coefficient subgrid-scale model for large-eddy simulation of turbulent scalar transport in complex geometries. Technical report, Center for Turbulence Research, Annual Research Briefs, 2007.

[125] L. L. Yuan, R. L. Street, and J. H. Ferziger. Large-eddy simulation of a round jet in crossflow. *Journal of Fluid Mechanics*, 379:71–104, 1999.